STANDARDS, EMERGENCE, AND COMPLEX OUTCOMES

Standards, Emergence, and Complex Outcomes redefines how we think about standards, framing them as interfaces that govern interactions and connect causes to their effects. Expanding beyond traditional technical and geopolitical discussions, Garcia introduces a fresh theoretical perspective that positions standards as central to understanding complexity. From food safety and workplace regulations to standardized parts in manufacturing and our reliance on technology, standards shape nearly every aspect of modern life.

By proposing a multidimensional approach, Garcia argues that standards are the mechanisms driving change, complexity is the backdrop for this change, emergence is the process by which change unfolds, and evolution is its ultimate outcome. Each chapter explores key elements of complexity theory and supports them with compelling historical and empirical examples, providing a richly detailed narrative that deepens understanding.

This book is an essential resource for academics and students across the social sciences, policymakers shaping standards-dependent fields, and professionals in industries where standards dictate progress. Whether you're exploring the intersections of technology, governance, and complexity or looking to better navigate the systems shaping our world, this book offers vital insights into the pivotal role of standards in our increasingly interconnected society.

D. Linda Garcia is Professor Emeritus at Georgetown University, where she taught courses on Technology and Society, Networks and the Creative Process, Global Standards, Networks and International Development, and The Networked Economy. She also worked for many years at the Congressional Office of Technology Assessment (OTA), which conducted bipartisan, interdisciplinary research assessing advancing technologies to determine how to maximize their benefits while minimizing their negative consequences. Among the studies that she contributed to were those having to do with transportation, acid rain, radioactive waste, educational technologies, and telecommunication and computer technologies.

STANDARDS, EMERGENCE, AND COMPLEX OUTCOMES

The Missing Link between Cause and Effect

D. Linda Garcia

Routledge
Taylor & Francis Group

LONDON AND NEW YORK

Designed cover image: zhengshun tang / Getty Images

First published 2026
by Routledge
4 Park Square, Milton Park, Abingdon, Oxon OX14 4RN

and by Routledge
605 Third Avenue, New York, NY 10158

Routledge is an imprint of the Taylor & Francis Group, an Informa business

British Library Cataloguing-in-Publication Data
A catalog record for this book is available from the British Library

Library of Congress Cataloging-in-Publication Data
Names: Garcia, D. Linda author
Title: Standards, emergence, and complex outcomes : the missing link between cause
and effect / D. Linda Garcia.
Description: Abingdon, Oxon ; New York, NY : Routledge, 2025. | Includes bibliographical
references and index.
Identifiers: LCCN 2024058596 (print) | LCCN 2024058597 (ebook) | ISBN 9781032721101 hbk |
ISBN 9781032721064 pbk | ISBN 9781032721125 ebk
Subjects: LCSH: Social sciences | Causation
Classification: LCC H61 .G295 2025 (print) | LCC H61 (ebook) | DDC 300.1--dc23/eng/20250429
LC record available at https://lccn.loc.gov/2024058596
LC ebook record available at https://lccn.loc.gov/2024058597

ISBN: 978-1-032-72110-1 (hbk)
ISBN: 978-1-032-72106-4 (pbk)
ISBN: 978-1-032-72112-5 (ebk)

DOI: 10.4324/9781032721125

Typeset in Times New Roman
by KnowledgeWorks Global Ltd.

For my mother, who challenged me to leave the world
in a better place than I found it.
For my husband, Brock, my greatest source of encouragement.
For Stephen, the best son anyone could ask for.
For my grandchildren, Sophie and Bennet, so qualified
to take up the torch.

CONTENTS

ACKNOWLEDGEMENTS

Having spent 20 years at the Office of Technology Assessment and 20 years at Georgetown University, I have had the priceless opportunity of engaging with scholars and thinkers from all walks of life. These encounters have led me to view the world in an interdisciplinary, multifaceted way, a perspective that is essential for understanding standards in our complex environment. For this I am very grateful. Special thanks in this regard go to the following: John Andelin, Johannes Bauer, Daniel Bell, Stanley Besen, Carl Cargill, Joseph Coates, Melvin Eggers, Louis Edinger, Sherry Emery, Dieter Ernst, Kai Jacobs, Robert Lekachman, David Lightfoot, Amy Kunz, Abbe Mowshowitz, Michael Nentwick, Eli Noam, James E. Price, Jim Turner, Ellen Surles, Michael Spring, George Venn, and Rick Weingarten. As importantly, I benefited and learned a great deal from my conversations and interactions with my Georgetown students, especially Kelsey Burns, Lu Hao, John Hanachek, Madhura Kale, Bethany Likely, and Ellen Surles. The curriculum development for my course on standards was funded, in part, by a grant from the National Institute of Standards and research grants from Georgetown University.

I am also thankful for the readers who provided comments on my chapters, especially those of John Alic, Jonathan Aronson, Lois Barry, Marjorie Blumenthal, David Clark, Robert-Cook-Deagan, Stephen Garcia, Kai Jacobs, Ken Kretchmer, Mark McCarthy, Eli Noam, Younie Sou, Michal Spring, George Venn. My grandson, Bennett Garcia, has been especially helpful, linking me to diverse aspects of the literature in computer science and the philosophy of science, as well as introducing me to the parallels between my work and the world of mathematics. I have gained greatly from the insights of all of these contributors. Especially instructive were readers who questioned my consideration of standards as both social as well as technical phenomena. I hope I have provided an adequate explanation of this approach by documenting the intellectual journeys that led me along this path.

I am overflowing with gratitude for all the wonderful people in my life. First of all, I want to thank my son Stephen, who—as mentioned above—first evoked my interest in standards while reading *What Do People Do All Day* to him as a child. We continue to share our intellectual interests, especially when it comes to networks and complexity. At the top of my list of supporters is my husband, legendary environmentalist Brock Evans, who has been steadfast in his encouragement in what he has described as my intellectual autobiography. Especially satisfying has been our shared love of history, which has led to many engaging discussions germane to my research. I also owe a great debt to my friend and caregiver, Laurie Stone, who has provided the support and safe space for me to spend my time engaged in thoughtful bliss. My very special thanks go to Dr. Geraldine Gardner and Dr. Emma Pereis, who, on discovering a problem with my heart, have done everything in their power to keep it running well. And, oh yes, I am blessed with my wonderful neighbors and friends—Wes, Jen, Don, Lois, Juanette, Mary, Margaret, Vito, Moira, and Ryan—who have always been there for me. To be so blessed!

When the standards have been set, things are tested and weighed. And the work of philosophy is just this: To examine and uphold the standards, but the work of a truly good person is in using those standards when they know them.

Epictetus, *Discourses*, 2.11.23–25

PREFACE

0.1 Synopsis

The world today is topsy-turvy. Borrowing from complexity theory, we might say that it is experiencing a phase transition, such that a crisis in one sphere of activity is spreading to others, threatening to bring chaos throughout. While some crises have been deliberately provoked by populist governments, others—such as the stock market collapse, the recent health crisis, and global warming—have occurred due to negative feedback loops stemming from the externalities associated with the sum of individual and group behaviors. Although social scientists typically address each of these problems individually, employing diverse disciplinary approaches, they have yet to develop a general analytical framework that would allow these situations to be analyzed in relation to one another. By drawing on complexity theory, this book, *Standards, Emergence, and Complex Outcomes: The Missing Link between Cause and Effect*, aims to do just that. While complexity theory characterizes the context in which social interactions occur, it is standards that play the lead role. As defined here, *standards are the interfaces that govern interactions*. As such, they determine who and what can interact, how interactions take place, and under what circumstances. Acting in this capacity, standards function as the building blocks of complexity.

Like all norms and institutions, standards emerge and coevolve through the interplay of social interactions, social institutions, and social norms, be they cultural, political, or economic. Conceived broadly, standards take the form of codes, protocols, norms, memes, signals, signs, laws, roles, etc. Viewed in terms of complex adaptive systems, standards provide the rules and mechanisms for emergence and social change. By interconnecting diverse actors and information, standards generate new capabilities and attributes, as well as a roadmap for actors, allowing them to enhance their fitness and rise up the fitness landscape. As they do so, the landscape, and the standards governing it, increase in complexity as well, raising the bar for all players. The theme that emerges from this effort is the critical role that standards play in the evolution of complexity and, with it, the evolution of life itself.

DOI: 10.4324/9781032721125-1

0.2 Encountering Standards Early On

We grow up with standards, so we take them for granted. Consider the books we read to our children. They serve not only to entertain but also to instill confidence by providing a world-view in which children can identify themselves, their relationship to others, and their places in the world. Years ago, I enjoyed reading Richard Scarry's book, *What Do People Do All Day* (1968), to my young son Stephen. I found the book to be a wonderful way of introducing him to his surroundings, demystifying the unfamiliar, and providing him with a sense of constancy.

Revisiting Scarry's book, I see it as a telling metaphor for the universe of standards. For, just as the tales in *What Do People Do All Day* clarify the roles, relationships, and norms of the fictitious place called Busytown, so today's everyday standards—be they technological, biological, economic, behavioral, or spiritual—provide a roadmap for negotiating our way in and around an uncertain world. A visit to Busytown can show us how.

As depicted by Scarry in his delightful, colorful illustrations, Busytown is a small community, like many others, populated by a teeming assortment of diverse animals engaged in a multitude of activities. Included among these characters, for example, are Mayor Fox, Farmer Alfalfa, Grocer Cat, Nurse Nelly, Stitches the Tailor, and Smokey the Fireman, to name but a few. Each character can be seen participating in some undertaking that serves to identify and illustrate his or her role in the Busytown community. Once these characters have been introduced, Scarry shows how, and to whom, they are related. In stories such as "Building a new house", "Mailing a letter", and "A voyage on a ship", we see all the people and artifacts that must come together in a given sequence so that such endeavors can be executed successfully. One has to wonder, given all the hustle and bustle, how do the citizens of Busytown manage to find their appropriate others? How do they figure out with whom they should interact? How do they locate the right tools to help them carry out their tasks? *The answer—it's all about standards!*

0.3 The Role of Standards

Like Scarry's account of Busytown, this book, *Standards: The Building Blocks of Complexity,* tells stories about standards, and the role they play in generating complexity and channeling complex outcomes. As defined herein, *standards are the interfaces that govern all interactions.* Accordingly, they might determine the mode of interactions, define the conditions under which interactions take place, and/or signify the appropriateness of interactions. As such, standards are the building blocks of the natural world and human society, as well as the glue that holds everything together. For, in any given context, standards constitute an agreed upon set of meanings, scripts, memes, codes, signals, algorithms, and rules that guide behavior and govern interactions. Embodying critical information in a highly compressed and abbreviated format, they greatly simplify the complex environment. Signaling opportunities and constraining choices, standards make possible cooperation and coordinated behavior.

Notwithstanding their critical importance, standards do not draw much attention. They typically reside in the background and only come to light when they are contested, dysfunctional, or absent altogether. Like most people, I was never particularly curious about standards. But once engaged, I saw standards everywhere. In this book, I explore their role in fostering complexity.

0.4   Searching for the Link between Cause and Effect

How did this happen? Years ago, I worked for the Congressional Office of Technology Assessment (OTA). We called it "Congress' own think tank." Our mission was to analyze upcoming technologies and assess their potential impacts, identifying what might be done to enhance their benefits and ameliorate their negative consequences. We worked in teams of diverse individuals so that we might benefit from the interaction among our disciplines as well as our unique thought processes. Much as complexity theory suggests, working through our different perspectives, we transcended the ordinary and emerged at a higher-level of understanding.

The major challenge facing us at OTA was developing an approach that linked cause and effect between technologies and their impacts. While there were numerous studies that sought to identify the impact of technology on society, there were far fewer that identified the actual means and mechanisms by which such change takes place (Misa,1992,1988; Beckwith, 1989). As problematic, although scholars had analyzed technology from a variety of angles, they had yet to develop the type of unified perspective necessary for advancement of the field (Misa,1992, 1988; Beckwith, 1989; Staudenmaier, 1999). Focusing primarily on the macro-level, some viewed technology as an independent variable, which essentially drives the course of human history (see Innis, 1951; Ellul, 1964; Beniger, 1989). In contrast, others conceived of technology as a dependent variable, the nature of which is determined through negotiations among relevant stakeholders in pursuit of their agendas (Bijker et al., 1987). In turn, actor network theorists, such as Bruno Latour, view technology as part of the equation. They conceive of technology as an "actant"—that is to say, an agency-endowed endogenous variable—and identify power in all its manifestations as the major determinant of outcomes.

Unfortunately, efforts to reconcile these diverse viewpoints have not been especially successful. They have been hindered by disciplinary divides, as well as by the ideological gap between technology determinists and social constructivists. While social constructivists claim that determinists lack realistic models of science and technology, technology determinists, such as Langdon Winner (1977; 2020), critique social constructivists for failing to take power and values into account. Equally problematic in precluding a common perspective is that scholars tend to focus on different levels of analysis (Misa, 1992; 1988, 3–12).

Through the course of my research at OTA, it became clear to me that technological outcomes depend not solely on the nature of a technology, as technology determinists propose. Nor can they be totally explained by the stakeholders involved in technological decision-making. Instead, outcomes are determined, to a large extent, by intervening and interconnecting variables—that is to say, the specific context and conditions under which technologies are deployed and diffused. I maintain that the most promising way of incorporating these variables is to characterize and formalize them as standards, and to analyze their evolution in the context of complexity analysis and emergence.

0.5   Standards Come to Light

When first asked by Congress to undertake a study of standards and the US standardization process, I was disappointed with my assignment. What could be interesting about standards, I asked myself? I knew that, when people are questioned about the subject, a typical response is: "My eyes glaze over."

I was in for a great surprise! My OTA standards project, *Global Standards: Building Blocks for the Future* (1992*),* turned out to be an exercise in investigative journalism—deep throats and all! For example, boxes of materials relating to standard setting were delivered anonymously to my office. I also received off-the-record calls, describing the misdoings and misdeeds of members of the standards setting community. Meanwhile, at standards gatherings, my staff and I were counseled by standards players not to talk to so-and-so, who was described as a "sleaze ball", or to some other person, who was referred to as a "scum bag". It was thus that I learned that standards were not just about technology at all!

0.6   Standards as Interfaces

It was during my work at the OTA that I determined that the best way to relate technology and society was to focus analyses on the *interfaces* that govern interactions and link technology and society. *These interfaces are, in fact, standards.* By defining standards abstractly, as interfaces, we can subsume both the standard rules that bind organizations together as well as the standard products, currencies, and prices that unify markets. In fact, viewed generically as interfaces, *standards might be said to take a variety of forms, including signs, symbols, codes, grammar, algorithms, memes, behavioral rules, and stereotypes, to name but a few*. Hence, employing the same conceptual framework, standards are amenable to analyses of all phenomena such that they are inherently interdisciplinary.

0.7   The Demise of OTA

Notwithstanding OTA's highly acclaimed nonpartisan research, Republicans in the 101ˢᵗ Congress decided to eliminate its funding.[1] Defending this position, Republicans claimed they had all the answers they needed in their *Contract for America*. The last thing they wanted was for OTA to disrupt their agenda by raising a lot of unnecessary and provocative questions.

The demise of OTA was a major loss. For twenty years, Congress had turned to our agency to analyze the role of advancing technologies. Although prediction comes with risks, our assessments were not only prescient; they were also very useful. When thinking about my years at OTA, I wonder what we might have accomplished had we been able to examine questions relating to social media, digital currencies, health pandemics, climate change, artificial intelligence, and space exploration. What made my job so compelling was the diverse subject matter with which we engaged and the joy of delving into unexplored and unresolved puzzles. Among the projects I oversaw during my twenty years at OTA were those relating to privacy issues, intellectual property rights, the future of communications, global trade and aid, electronic commerce, as well as the US stake in global standards setting. Having become hooked on the subject of standards, I was fortunate to be able to pursue my interest as I moved on from OTA to Georgetown University.

0.8   Teaching Standards

When I left OTA in 1995, I had the great pleasure of directing the recently established master's program, Communication, Culture, and Technology (CCT), at Georgetown University. Like OTA, CCT is an interdisciplinary program that thrives on viewing technology-related questions and issues from diverse disciplinary perspectives. It was in this interdisciplinary environment that I was able not only to pursue my growing interest in standards but also to expand the scope of my inquiries and to link them to the burgeoning fields of networks and complexity. In so doing, I greatly benefited from the growing literature on both subjects, which became the theoretical grounding for my courses and subsequent research efforts.

To support my teaching on standards, I was thankful to receive a grant from the National Institute of Standards and Technology (NIST). Looking for course materials to share with my students, I found that there were few standards courses available within the university community. Nor were there many refereed publications upon which to build. Under the circumstances, the number of students taking formal classes learning about standards was minimal. Responding to this situation, the NIST provided me with a grant to develop a standards course for students at the college and graduate school levels.

0.9   Formalizing Standards in Relation to Complexity

It was not long into my time at CCT that my interest in standards converged with my interest in complexity, enabling me to formalize the role of standards in generating the emergence of new kinds of interactions, giving rise to greater fitness and higher-level fitness landscapes[2]. It is this process that I follow in tracing the narratives laid out in this volume. The focus of my inquiries is on interactions and the standards interfaces that govern them, be they related to technologies, cells, animals, plants, or people. I argue that diverse agents, operating in complex environments, enhance their fitness, and work their way up the fitness hierarchy by organizing and interacting in new ways based on access to reconfigured connections and new information. In so doing, new types of agents and actions emerge, in accordance with new standards, that lead to greater fitness for the agents as well as for the fitness landscape as a whole. It is this process of emergence, together with new standardized blueprints facilitating adaptation, that constitutes the mechanisms driving social change and the process of evolution over the long-term.

0.10   Motivation for the Book

Ironically, the lack of attention to standards has come at a time when standards are becoming ever more critical, given our increasingly complex and interconnected world. Consider, for instance, the banking system where new forms of currency are being created faster than we can develop standards governing their usage. The food industry is also increasingly complex. Notwithstanding labels denoting organic foods, gluten-free foods, genetically non-modified foods, and antibiotic-free foods, we have been at a loss when trying to customize our diets to our needs. We have all read, with some horror, about adulterated baby milk in China and contaminated chicken products in the United States. Especially troubling to me was the discovery that the "kibbles" I had been feeding my dog, Sweet Pea, contained ground-up animal bones from most any kind of animal—even roadkill, intestines, and other organs, as well as scraps swept up from the food-processing floor. Most disturbing, however, is not simply the failure of standard setting to keep pace with our increasingly complex environment. Equally problematic, from my perspective, has been the effort by those in the highest ranks of government to gain their private ends by undermining the standards and norms reflective of our society. What, for instance, should we make of Trump's call for the termination of our Constitution or the prevailing chaos in the US House of Representatives?

The idea for this book emerged from these concerns. It is based on the belief that the tepid interest in standards is due to the narrow way in which we have typically conceived of them, as well as a lack of attention to their system-wide social, economic, political, and cultural impacts. Today, most standards efforts and analyses focus almost entirely on the technical and economic aspects of standards. No doubt these topics are essential to our understanding of standards and standards processes. And, in fact, it was my work on technology-related subjects that led me to investigate the role of standards in broader contexts. Exploring different venues, we can uncover the full range of questions to which standards give rise. By characterizing standards and standards issues more broadly, and in the context of complexity, we can better assess the importance of standards and their role in determining much of what takes place in our highly complex world. As we shall discover, *standards are, in fact, the missing link between our intentions and their outcomes, between our technologies and their impacts.*

By making the role of standards explicit, I hope to highlight what is at stake for us as individuals, collectivities, and global actors. It is with this in mind that I have developed this manuscript. My intent is to provide the basis for a general theory of standards, there being no such thing at present. For a foundation, I build on complexity and network theory. Living, today, in a complex society that is increasingly comprised of networks and interactions, this would seem to be an appropriate place to start. If there is an overall theme that emerges from this effort, it is the critical role that standards play in the emergence of complexity and, with it, the evolution of the universe itself. I contend that diverse interactions are the "proving ground" of complexity and that standards are critical to these processes insofar as they channel and govern them.

0.11 Methodology

My academic career provided me with broad brushes for analyzing social, economic, cultural, and political phenomena. After graduating from Syracuse University in 1969, I started my graduate studies at Columbia University in the late Sixties and early Seventies, a time when many scholars were preoccupied with "The Big Picture" and the search for what—during pre-war Europe—had gone so utterly wrong. Academics of this period wanted to set things right, but to do so, they needed to understand, or at least hypothesize, the relationship between causes and effects. The result was the outpouring of a number of narrative case studies based on historical data from which many useful inferences were drawn.

It was in this milieu that I did my graduate work, first at Columbia University's School of International Affairs and later in its Political Science Department. My focus was comparative government and politics. It was from here that I inherited the broad-based intellectual tradition of my professors. Among the most important to me were Otto Kirchheimer, Bela Balassa, Juan Linz, Albert Hirschman, Philip Mosely, Robert Lekachman, Todd LaPorte, senior, and Lewis Edinger. Thanks to Professor Abbe Mowshowitz,[3] who served as my promoteur, I completed my PhD thirty years later at the University of Amsterdam, having taken a break to raise my son Stephen and to work at the Office of Technology Assessment.

Following the lead of my former professors and mentors, I continue to employ a narrative, historical approach in my research, analyzing evolving sequences of events and the interactions that engendered them. But, in addition to identifying the paths linking cause and effect, I focus on, and formalize, the intervening variables that skew the trajectory of causal actions and lead to indeterminate, complex outcomes. *I define these intervening variables as standards, which I argue are the building blocks of social structure (understood in the broadest terms).* When these standards are altered to a significant degree, they generate a phase transition, such that new relationships and possibilities emerge together with innovative outcomes.

0.12   An Overview of the Book

Building on the tradition of historical sociology and my training in comparative government, the chapters in this book analyze diverse sequences of historical events and circumstances in light of our expanding understanding of complexity. Employing this framework in each case, a significant conclusion emerges. Standards play a critical role not only in determining technology outcomes but also—and far more importantly—in channeling our evolutionary future. The chapters supporting this conclusion are described below.

The book is divided into four parts. The Preface and the Introduction constitute Part I, in which I raise my research questions and the theoretical framework and methodology I employ to address them. Part II, Chapters 2 through 6, details the role of standards in relationship to complexity by employing empirical cases to illustrate and explain the various attributes of complexity theory. Part III, Chapters 7 through 11, focuses on the role standards play in generating the complex structures that function as platforms for emergence. Part Four—the conclusionary chapter, "Putting the Pieces Together"— summarizes the book's conclusions and its implications for the social sciences. Although each chapter can stand alone, they build upon one another; so that when viewed together and in sequence, they provide a coherent conceptualization of complexity. The content of each chapter is summarized below.

Chapter 1. *The Introduction.* This chapter provides the conceptual framework and methodology that guides the analyses in the following chapters. Claiming interactions to be the appropriate unit of analysis, the chapter defines standards as the interfaces governing interactions and explains how, when taken together, they create platforms upon which adaptation and innovation emerge. So conceived, standards constitute the missing link between cause and effect. Accordingly, the chapter proposes that (1) *standards* are the mechanisms of change; (2) *complexity* is the context in which change takes place; (3) *emergence* is the outcome of the process, (4) leading reiteratively to greater complexity. Illustrating how this occurs, each chapter characterizes specific aspects of complexity theory, describes the role of standards relative to them, and corroborates each theoretical construct with empirical, historical accounts.

Chapter 2. *Evolving Fitness—How the West Was Won.* It introduces the concepts of fitness and fitness landscapes, which are fundamental to understanding how systems adapt in response to altered environments. This chapter documents how diverse standards provided American pioneers the stepping stones needed to ascend the fitness landscape and emerge in the "promised land" of Oregon. Detailing the process by which pioneers made their way to the US west, the chapter accounts for both how and why standards lead to the evolution of emergent outcomes. It illustrates, moreover, that outcomes are not necessarily positive or neutral; rather, they give rise to winners and losers, depending on the standards governing interactions.

Chapter 3. *Norms and Emergence in Social Settings.* It details how emergence takes place when two or more entities are integrated together, based on converging standards, such that the outcome of their amalgamation is not only decidedly different from, but also more complex than, the sum of their parts. Comparing a number of diverse empirical cases, the chapter illustrates how, in each instance, standards—taking the form of norms, conventions, day-to-day practices, symbols, protocols, codes, etc.—engender emergent processes and structures that give rise to greater complexity and fitness at sequentially higher levels of the fitness landscape.

Chapter 4. *Platforms—the Springboards for Evolution.* Platforms can best be conceived of as hubs where networks of diverse interactions converge around standards in a multiplexed fashion when actors/actants, as well as their purposes and functions, align. Comprised of a densely connected center linking outward to their external environments, platforms constitute *small worlds* where new information and resources are agglomerated and new standards and practices are generated, so that adaptation and emergence up the fitness landscape, take place. Focusing on the role of platforms, the chapter compares cases in the past to those at present. It argues that today's digital platforms have become so all-encompassing and densely interconnected that they function less and less as small worlds.

Chapter 5. *Standards and Phase Transitions in the Middle Ages.* This chapter introduces the concept of phase transitions. Building on complexity theory, it traces the history of the Middle Ages, detailing how, and why, standards-driven emergent processes gave rise to major transformations and to what effect. Focusing on the role of interactive ties, the chapter demonstrates how radical changes in the social structure occur periodically due to phase transitions brought about by rigid standards configurations that prevent adaptation to changing circumstances. In the process, prevailing powers are undermined, and activities are reestablished and reconfigured around new standards, generating thereby greater complexity, higher fitness, and new opportunities for some, while foreclosing them for others.

Chapter 6. *Standards and Evolution in a Complex World.* Evolution is about survival. It is the means by which actors/agents/genes/artifacts seek—through the process of emergence—to enhance their fitness in the context of a changing environment. Central to this undertaking is a procedure whereby members of a population are selected to advance up a given hierarchy, and be replicated at a higher fitness level in accordance with a set of fitness criteria—defined here as standards. Although present-day accounts of evolution incorporate many aspects of this characterization, a unified perspective, which encompasses all aspects of the world as we know it, has yet to be achieved. Disagreements stem not only from disciplinary divisions but also from distinct methodological approaches, especially those relating to reductionism[4] and the appropriate unit of analysis. By viewing evolution through the lens of complexity theory and the role of standards, this chapter aims to transcend these divisions so as to provide a more integrated, and standards-based, rendition of evolution.

Chapter 7. *Monetary Outcomes—Winners, Losers, and the Standards That Determine Them.* This chapter challenges economists' claims that money, operating as an "invisible hand", is neutral in its impacts. It defines money as an emergent phenomenon that arises from, and is determined by, interactions within the social structure in which money is embedded. Money is conceived as an intersubjective, social phenomenon that serves as a standard of value. Having agency of its own, money is not neutral. It determines winners and losers as well as long-term social structures. To illustrate the impact of money, the chapter references the case of the Robber Barons in the late 18th century United States.

Chapter 8. *Standards, Modularity, and Innovation.* It focuses on the processes by which innovation comes about. Demonstrating the universal nature of these processes, the chapter compares and details how innovations occur with respect to technologies, biological systems, organizations, and social structures. In contrast to many economists who have argued that standards retard innovation because they become locked in due to their first-mover advantages, as well as the externalities associated with them, the chapter argues that such need not be the case. Laying out that possibility, it documents how standardized, modular components can be reconfigured to fashion not only new and greatly improved innovative technologies but also transformative societal innovations. Moreover, much as complexity theory predicts, the chapter illustrates how innovation is enhanced when new information and new types of interactions, are integrated within the process.

Chapter 9. *How Standards Engender Trust.* It documents both the importance of trust as well as its decline in today's postmodern environment. After considering various conceptualizations of trust, the chapter characterizes trust as an emergent, relational phenomenon. Like money, trust is intersubjective, arising in the context of standardized interactions and rules of behavior. In accounting for the present-day decline of trust, the chapter points to the increasing disembeddedness of human interactions, and with it a decline of common standards. It argues that regaining trusting relationships will require the re-embedding of societal interactions, as well as the recalibration of standards so as to generate revitalized platforms of trust.

Chapter 10. *Crafting Identity Platforms with Standard Memes, and Symbols.* It builds on role theory and the work of Erik Goffman and Randall Collins to characterize identities as platforms comprised of standardized roles that emerge from within the existing reservoir of existing cultural memes and social symbols. It shows how roles serve as platformed agglomerations of "identities", linked together by the standards that both define and govern them. Like all small worlds, these platforms constitute major sources of power. Examining the role identity platforms have played at the national level, both in promoting and undermining political power, the chapter traces the political impact of role identities from the time of Louis IVX to that of Napoleon Buonaparte.

Chapter 11. *The Artist as Standards Bearer.* It characterizes the role of artists as standards bearers who, through their works, employ their agency to influence activities in broader social and political contexts. As importantly, it describes how artists, by embedding their ideas within their artwork, serve as standards bearers, providing seed corn in the form of cultural codes and memes that solidify their ideas, passing them on to future generations. Because artists' influence is especially compelling during periods of upheaval, changes in artistic styles have typically corresponded to periods associated with major phase transitions. Three cases illustrate this phenomenon: the periods of French neoclassical art, romantic art in France and Germany, and *avant-garde* art in the Weimar Republic.

Chapter 12, *Standards and Complexity—Piecing the Puzzle Together.* This—the concluding chapter—identifies standards as the mechanisms driving complexity as well as the linchpin linking cause and effect in complex systems. To this end, it characterizes the role that standards play with respect to each of the attributes associated with complexity, as identified herein. It illustrates how, in each instance, standards not only determine complex outcomes, but also link outcomes together so as to form a coherent whole. The analysis provides a basis not only for explaining outcomes, but also for identifying leverage points within complex systems that can be employed to shape them. The discussion is based on both the theoretical literature relating to complexity as well as the empirical analyses put forward in the preceding chapters.

Notes

1 The legislation establishing OTA is still on the books, it has just not been refunded. There have been numerous calls for reinstating OTA, which have yet to garner enough support to refund it. However, in light of the increasing number of the scientific and technological issues that we face as a society the idea is increasingly in the air. See, for instance, https://www.brennancenter.org/our-work/analysis-opinion/whether-supreme-court-rolls-back-agency-authority-congress-needs-more/

2 The concepts of fitness levels, and fitness landscapes were developed by Sewall Wright, a population geneticist. Generally speaking, fitness refers to the extent to which an entity conforms to the demands of its environment. A fitness landscape refers to the sum of the points in an environment that define the criteria for fitness such that each point represents unique criteria.

3 Abbe Mowshowitz is a mathematician, a computer scientist, as well as a social scientist. He is also a Professor at City University of New York. For me, Abbe was a mentor and an inspiration, encouraging me to pursue interdisciplinary analysis in the area of technology and society, and to complete my doctoral degree at the University of Amsterdam. I am forever grateful. See Mowshowitz (2002).

4 Reductionist posits that, to fundamentally understand a phenomenon, one must begin by analyzing its individual parts, and then, on that basis, deduce the nature of higher-level structures. Their approach assumes that events, when carried out in a process, are linear, so that outcomes can be anticipated and explained by simply "adding up" the steps in their development (Kontopoulos, 1993; Kauffman, 1996; Bunge, 2003, 23).

References

Beckwith, Guy V. (1989) "Science, Technology, and Society: Consideration of Method," *Science, Technology, and Human Values* 14 (4), 323–339.

Beniger, James (1989) *The Control Revolution: Technological and Economic Origins of the Information Society*, Cambridge, MA: Harvard University Press.

Bijker, Wiebe, et al. (1987) *The Social Construction of Technological Systems: New Direction in the Sociology and History of Technology*, Cambridge, MA: MIT Press.

Bunge, Mario (2003) *Emergence and Convergence: Qualitative Novelty and the Unity of Knowledge*, Toronto, CA: University of Toronto Press.

Ellul, Jacques (1964) *The Technologic Society*, New York, NY: Vintage Books.

Innis, Harold A. (1951) *The Bias of Communication*, Toronto, CA, University of Toronto Press.

Kauffman, Stuart (1996) *At Home in the Universe: The Seach for the Laws of Self-Organization and Complexity*, New York, NY: Oxford University Press.

Kontopoulos, Kyriankos, M. (1993) *The Logics of Social Structure*, Cambridge, UK: Cambridge University Press.

Misa, Thomas (1988) "Theories of Technological Change: Parameters and Purposes," *Science, Technology, and Human Values* 1 (1), 3–12.

Misa, Thomas (1992) "How Machines Make History and How Historians (and Others) Help Them Do So," *Science, Technology and Human Values* 13, # 3 +4, Summer and Autumn 308–333.

Mowshowitz, Abbe (2002) *Virtual Organization: Towards a Theory of Societal Transformation Stimulated by Information Technology*, West Port, CT: Quorum Books.

Staudenmaier, John. ed (1999) "Technology and Culture," Single Issue, *The International Quarterly of the Society for the History of Technology* 40 (1).

Winner, Langdon (1977) *Autonomous Technology: Technics Out of Control as a Theme in Political Thought*, Cambridge, MA: MIT Press.

Winner, Langdon (2020) *The Whale and the Reactor: A Search for Limits in an Age of High Technology*, 2nd edition, Chicago, IL: University of Chicago Press.

PART I

Motivation and Conceptualization

1

CONCEPTUALIZING THE PROBLEM

1.1 The Task at Hand

The social historian, Charles Tilly, urged that when doing analyses of social processes, we should focus on interpersonal transactions and social ties.[1] As he pointed out: "transactions compound into identities, create and form social boundaries, and accumulate into durable social ties". The role of the analyst, according to Tilly, is to identify "recurrent causal, [relational] mechanisms that produce alterations among social sites" (Tilly, 2016, 17–18).

In this book, *Standards, Emergence, and Complex Outcomes: The Missing Link between Cause and Effect*, I follow Tilly in asserting that interactions should be the basic focus of analysis.[2] However, I believe that Tilly's characterization of the social mechanisms,[3] according to which interactions occur, needs to be further theorized. As Tilly maintains, social mechanisms can be identified by searching for recurrent patterns of human interactions in diverse circumstances. While such an approach addresses how recurrent circumstances and behaviors lead to events, it has much less to say about the underlying means and modes of interactions. To understand both the how and why of interactions, we need to look deeper—at lower levels—to consider how interactions are themselves governed. Such an undertaking requires an investigation into the role of standards and how they bring about evolutionary change. This is the task addressed in this book.

I have been studying standards for over 20 years both at the Congressional Office of Technology Assessment) and at Georgetown University.[4] Based on my research, I contend that standards have agency in their own right. In fact, standards constitute a critical link between cause and effect. Thus, *I define standards as mechanisms (social and otherwise) that govern interactions.* Embodying values and norms as well as affordances and constraints, standards determine who and what can interact, by what means, and under what circumstances. Serving, both as signals and boundaries, standards provide the ties and channel the pathways that allow individuals to connect to one another in groups, networks, economies, and nation-states. As importantly, by codifying interactions, standards reduce uncertainty as well as harness complexity.

DOI: 10.4324/9781032721125-3

In this introductory chapter, I lay out a picture of the standards universe. In it, I characterize the importance of standards, differentiate between them, specify how standards function as mechanisms governing interactions, explain how standards discriminate between winners and losers, and describe how they instigate social change. As importantly, I spell out how changing standards configurations generate phase transitions[5] that lead to major social transformations requiring novel standards and unconventional modes of behavior. One need only consider, for example, how the decline of industrial standards accompanied, and accelerated, the shift to new types of business arrangements in the post-industrial environment (OTA, 1994). Today, platform standards are likewise increasingly being used to reconfigure business relationships (van Schewick, 2012; Parker et al., 2017).

1.2 Standards, Standards, Everywhere

Standards are ubiquitous. Food and drug providers must comply with health standards; cars use standardized interchangeable parts; workplaces have safety standards; clothing comes in standard sizes; cells operate in accordance with standard codes and signals; workers perform standardized roles; groups form around standardized rituals; telephones have standardized interfaces; and bedsheets are sized to fit standardized mattresses. Even our lives have become standardized through our reliance on technology (Winner, 2020). Hence, what roles standards play and how they are selected and adopted are of considerable importance.

1.2.1 Follow the Leader

To appreciate the role of standards, consider the origin of the word. Initially, the word "standard" signified a flag or banner that was associated with a given leader and hence used to rally his troops in battle (Malone, 1932). Armies throughout history have made critical use of such standards. For example, the standards bearer, who was the most important person in the field, took the front-most position. Holding the flag high, he gave hope and heart to his compatriots. But were he to fall in battle and the enemy to appropriate the standard, it would signal defeat. Thus, it was that Prophet Isaiah predicted that the Assyrian attack against the Hebrews would fail with the "fainting" of the standards bearer (Isaiah, 1611, 18).

Standard signals continue to play such leadership roles today, even in the animal kingdom. Take slime mold, for instance. Instead of rallying warriors to battle, slime molds signal the presence of food, or lack thereof, via rhythmic pulses of cyclic adenosine monophosphate. When food is prevalent, individual slime molds forage on their own. However, when food is sparse, they cluster together and build a hard stalk, the top of which houses inert spore cells. When conditions improve and all active cells have died, the spore cells reemerge to take their place in a resurrected community (Holland, 2011, 13). Likewise, ants carry out collective tasks on behalf of their colonies. They facilitate collaboration by employing pheromones to signal the location of food sources, as well as to identify each ant's specific task. Honeybees, too, are social beings. Bee colonies also depend on standardized communications for survival. Chemical codes within the gnome determine the roles each bee plays within the hive, while coded communication dances signal the direction and distance of food sources, guiding bees as they interact for their benefit, as well as the benefit of their hives (Kirschner and Gerhart, 2005, 103–105).

1.2.2 Linking People and Things

Language and gestures play a similar role for humanity. Based on a common understanding, language provides the shared frame of reference and sense of reality that allow people to have intimate relationships and common goals. Similarly, cooperation among individuals engaged in interdependent activities is greatly facilitated when people do not act randomly, or on a trial-and-error basis, but rather conform to common expectations about socially constructed roles (Katz and Kahn, 1978; Biddle, 1979, 1986). Where languages are distinct and used in common, they signal the presence of a culturally based community (Anderson, 2016).

Such ties are ever present throughout society. Organizations, for example, gain greater resources and reduce their transaction costs when they adhere to common standards and procedures institutionalized in their environments. In so doing, organizations become standardized themselves, as the prevalence of bureaucratic forms attests. These developments bring to mind Max Weber's concept of the "Iron Cage", a metaphorical prison that people build for themselves when, in their efforts to maximize profits, they focus on being totally 'rational' in their choices and behavior (Weber, 2001 [1930]).

Technology is likewise furthered by standard specifications and protocols. These add value to system components by allowing them to interconnect and interoperate in a transparent and seamless fashion. In the process, standards not only generate significant externalities[6] but also provide a platform on top of which other technologies and activities can ride (Grewal, 2009, 2009; Johnson, 2011; Parker et al., 2016). As importantly, standards allow technologies to be decomposed into their components, which can then be mixed and matched to create new technologies (Arthur, 2009; Johnson, 2011; Kauffman, 2019).

Standards also serve as general identifiers. For example, in the case of trademarks, standards distinguish products and help people sort through extraneous information to make better choices. At the same time, standards can inhibit or block interconnections because they signal the boundaries that cordon off groups, processes, and components (Holland, 2014). In fact, as we increasingly witness today, when employed as stereotypes, standard tropes may contribute to divisiveness rather than cooperation (Buchanan, 2007, 162–166).

1.3 High Stake Contests

The stakes entailed in standards choices are, therefore, inordinately high. Businesses can rise or fall based on technical, legal, and business process standards. Exchange standards, monetary standards, and quality standards provide the trust without which trade could not take place. Wars are fought on behalf of standards, whether religious, political, or otherwise. At the same time, wars are fought in accordance with international standards of conflict, such as the Geneva Convention. As importantly, children learn how to speak and how to behave in the context of cultural standards, while adults find their places in the world based on societal standards. Even nature abides by standards, as in the case of fungi, which interconnect trees in a common network, often referred to as the "wood wide web" (Sheldrake, 2020, 9).

Notwithstanding the high stakes involved, we are typically unaware of standards because we take them for granted. Yet we recognize them all too well when they are absent and we need them or when they are not to our liking. But we ignore them at our peril. Just consider the fate of the man who reaches out to pet a dog even though its hackles are raised. Similarly, what would you say are the prospects of the student who dresses inappropriately for a job interview? And one can only imagine the sorry sight of the driver who, in a hurry, runs a red light!

1.4   Winners and Losers

All the while, we must keep in mind that standards are not neutral. They typically favor some at the expense of others. This occurs because standards have a strategic value in that those who control a standard often control all the activities associated with it (Grewal, 2009; Wu, 2010). Notable, too, is that network standards increase in value the more they are adopted due to the growth of networks based on those standards and the externalities associated with them (Arthur, 1989; Shapiro and Varian, 1998; Rolfs, 2001; Grewal, 2009). The Internet standard, TCP/IP, is a prime case. In the early stages of the Internet's development, adopters and commercial providers were few. However, with the expansion of the network and network applications, businesses emerged to capitalize on the enhanced value of the TCP/IP standard (van Schewick, 2012; Garcia, 2016).

1.4.1   Westinghouse vs. Edison

Often, standards are at the center of battles between industry titans. Consider the first "standards war". At the turn of the century, George Westinghouse and Thomas Edison aggressively fought with each other to define the standard for electric current (McNichol, 2011). At stake for Edison was DC, the foundation of his electrical empire, whereas George Westinghouse's operation was based on AC. Edison went to great extremes to win, engaging in a shameful campaign that was designed to generate fear about AC's safety. For this purpose, he undertook several trumped-up, grisly experiments involving the electrocution of dogs, cows, and horses. Notwithstanding Edison's efforts, AC—which could travel further and more efficiently than DC—won the standards war (McNichol, 2011; Garcia, 2016).

1.4.2   The Browser Wars

Battles such as these continue today, and they are carried out just as fervently (Wu, 2010). Consider, for example, the browser wars in 1995 when Netscape Navigator dominated 90 percent of the browser market and 90 percent of the installed user base (Garcia, 2016). As such, Netscape Navigator threatened Microsoft's dominant position in the operating system market because it worked across multiple network platforms, allowing software developers to create software for any operating system (Ryan, 2010; Garcia, 2016). But, in the end, Microsoft overcame its rival by combining its browser, Internet Explorer, with its operating system and affixing it, annually, to the desktops of 50 million new computers. By offering its browser at no cost, Microsoft succeeded in making Microsoft Explorer the browser of choice (Garcia, 2016).

1.4.3   Sociocultural Battles

Standards battles can be just as intense in social and cultural realms. One need only recall a few of the cultural conflicts throughout American history (Hartman, 2015). The fierce battles in the 1920s between fundamentalists and liberals in the northern Protestant denominations stand out. Each side was convinced disaster awaited were the other side to win. As Barry Hankins describes in his book, *Jesus and Gin: Evangelicalism, the Roaring Twenties and Today's Culture Wars* (2010), the evangelical movement went dormant after the Scopes trial,[7] only to reemerge once again in the 1980s (Hankins, 2010; Stokes, 2011). The intense cultural conflicts dividing the US population today are just the latest evidence of these wars.

1.5   Definitions Matter

Given the high stakes entailed in standard choices, how we define standards has major implications. Definitions used in everyday speech are not very useful. Businesses, for example, characterize standards, and design their standard strategies, based on the type of standard at hand. If standards are to be set by a legislative body, as in the case of the net neutrality rules or the Digital Millennium Copyright ACT (DMCA), businesses might employ rhetoric to link their preferred standards to national goals (Garcia, 2005). If, on the other hand, standards are to be set in the marketplace, businesses will likely define them in ways that appeal to potential supporters and allies in the hope of tipping the standard battle in their favor.

Others have their own takes on standards. Government policymakers naturally view standards in terms of regulations. Economists focus on standards in terms of their interoperability, competitive effects, and market failures. Social scientists, in turn, analyze standards to determine how conflict, collective action, and integration might take place, or not. Anthropologists study the norms, values, signs, symbols, and literary tracks that constitute a common culture, while historians compare patterns of interactions and events that link people in specific times and places (Lachman, 2013).

Given these diverse perspectives, we need a generic, all-encompassing definition of standards that applies equally to the standard rules that bind organizations together, as well as the standard products, currencies, and prices that unify markets. Moreover, because standards are necessary for all types of interactions, a definition must accommodate all forms and formats of standardized communications—signs, symbols, codes, grammar, tropes, behavioral rules, and stereotypes, among others. One way to achieve this breadth is to define standards in the abstract.

1.6   Standards Defined

Following the recommendation of James Rosenau and Mary Durfee in their classic book, *Thinking Theory Thoroughly* (Rosenau and Durfee, 1999), I begin by asking: "What is a standard an instance of?" The answer constitutes the top rung of the ladder of abstraction. Surveying the standard's universe from this height, we see that standards are the *means, modes, and mechanisms* of interactions. Taken together, they constitute a platform for interaction, signaling the boundaries of diverse phenomena and negotiating the boundaries within, between, and among these phenomena Holland, 2014). In the process, they generate the rules, or protocols, to be followed so objects can interact.

With this in mind, I define standards as the *interfaces governing interactions, be they between individuals, machines, or elements of the natural world.* Standard interfaces govern who and what can interact, the mode of interaction, the conditions under which interactions take place, and the appropriateness of interactions. For example, connecting to the Internet requires the TCP/IP protocol; driving on highways, cars must meet national emission and safety standards; purchasing produce, consumers look for standard certifications.

1.7 The Power of Standards

Standards are essential to the social order. They provide the "mechanisms" that scientists in the fields of physics, chemistry, and—in some cases—biology often claim to be lacking in the social sciences. As importantly, standards constitute the "social facts"[8] that individuals and groups must contend with as they negotiate their way forward" (Demeulenaere, 2011a, 2011b; Brennan et al., 2013, 3)

1.7.1 The Missing Mechanisms[9]

Scientists in the "hard sciences" have often downplayed the analytic value of the social sciences, arguing that their analyses are vague, redundant, and without explanatory power. They claim that unlike physics, chemistry, and biology, the social sciences typically fail to identify the *mechanisms* linking cause and effect. On that basis, some argue that only the physical sciences can generate rigorous, predictive models of the world (Hoyningen-Huene and Wuketits, 1989; Sawyer, 2005; Kauffman, 2019). Touting this perspective, many natural scientists have, for example, argued that to get to the truth, there is only one way to go—that is, by sorting out the elements at the molecular level and, from there, generalizing about those at the top level (Hoyningen-Huene and Wuketits, 1989; Kauffman, 1995; Kauffman, 2019). As Elder-Vass points out, reductionist assertions,[10] such as this, strike at the very heart of the social sciences (Elder-Vass, 2010, 15–16). It is not surprising, then, that many social scientists have had a strong, negative reaction to the rising field of sociobiology[11] (Turner and Machalek, 2018).

1.7.2 Analytical Sociology

Even within the social sciences, there are many who criticize the amorphousness of the discipline's research efforts and call for a more rigorous methodology—one that specifies the mechanisms by which social change takes place.[12] They call their approach "analytical sociology". According to Demeulenaere, the aim of analytical sociology is to "clarify the basic epistemological, theoretical, and methodological principles fundamental to the development of sound description and explanation" (Demeulenaere, 2011a, 2011b, 12; see also Hedstrom, 2005; Elster, 2015).

Although a number of analytical sociologists focus on individuals as the unit of analysis, analytical sociologists avoid methodological individualist explanations.[13] Instead, they consider individuals to be social and therefore subject to their interactions with others as well as to social constraints. Hence, these sociologists agree that individuals are embedded in situations and environments that partially determine their choices[14] (Granovetter, 1985). Equally important, they deny the notion that individuals are motivated by instrumental goals alone. Rather, individuals are thought to make choices based on personal inclinations and normative considerations as well as self-interest (Elster, 2015). As Demeulenaere claims, these constraints and considerations can be thought of as *social facts*[15] insofar as they:

> are generated, triggered, produced, brought about, or 'caused' by individual actions which themselves are in some sense 'caused' or at least partly determined by the constraints presented by the social environment and situations in which such actions take place.
>
> *(Demeulenaere, 2011a, 2011b, 23)*

Taken together, these "social facts" constitute *social structure*.[16] The task for analytical sociologists, then, is to identify within the social structure critical actors, structures, their interrelationships, and the situations and environments in which they are embedded, so as to identify consistent, and explicit, *mechanisms* that account for social change (Hedstrom, 2005).

The emphasis on mechanisms, social structures, and social facts is what distinguishes analytical sociology from other subfields of sociology. However, notwithstanding their emphasis on mechanisms, these scholars have yet to identify a generic mechanism that can explain outcomes in all cases. In particular, they have failed to link to an ontology continuous with the natural sciences (Sperber, 2011, 75). As Bunge (2003) has described the problem:

> An adequate and general definition of the conditions for emergence is elusive, if not impossible, given the large variety of emergence mechanisms... we need different theories to account for widely different emergence mechanisms. This is why scientific explanations are specific: because mechanisms are specific. In other words, there are no all-encompassing explanations because there are no one-size-fits-all mechanisms. (40)

This book contests this position. As described below, it contends that standards are, in fact, a generic mechanism that generates emergent behavior in all cases of interaction.

1.7.3 Standards as Social Structure

Although in the modern era we have come to think about standards in technical terms, standards are first and foremost the building blocks of the social order—itself a network of standards-based networks (Kontopoulos, 1993; Sawyer, 2005; Beinhocker, 2006). For, in any given context, standards constitute an agreed-upon set of meanings, scripts, and rules that guide behavior and govern relationships. Embodying critical information in highly compressed and abbreviated formats, standards greatly simplify the environment. Signaling opportunities and constraints, they allow for cooperation and coordinated behavior to take place (Garcia et al., 2005). By providing an overarching, common point of reference, standards help integrate social systems. Most importantly, in today's environment, standards—serving as interfaces across boundaries and between and among actors—not only allow interconnection and feedback to take place; as importantly, they generate innovative, emergent behavior allowing for evolutionary adaptability. We can witness these possibilities today as we turn our attention to complexity theory and complex adaptive systems.

1.8 Complex Adaptive Systems

Standards play an exceptionally important role in complex adaptive systems.[17] In fact, it is by virtue of standards that emergence—a defining feature of complexity—takes place. Because complexity and standards are both interdisciplinary and interlinked phenomena, I examine the relationship between standards and social outcomes through the lens of complexity.

1.8.1 Key Attributes of Complexity

Although complexity analysis has yet to assume an all-encompassing body of theory, we can characterize complex adaptive systems by the attributes ascribed to them[18] (West, 2018, 22).

Surveying the literature on complexity, we see that complex adaptive systems encompass a number of independent, heterogeneous actors. Acting according to their own unique scripts and roles, these actors affect the behavior of all others in the system (Monge and Contractor, 2003). Complex systems are self-governing. That is to say, the result of their joint interactions is emergent such that the "whole is greater than the sum of their parts" (Kontopoulos, 1993). Changes and interactions generated from the micro level of the system give rise—with the help of standards—to "self-organization" up the fitness landscape so that outcomes at the macro level transcend lower-level actions and cannot—as in linear systems—be traced back to them (Kontopoulos, 1993; Bunge, 2003; Kauffman, 2008; Monge and Contractor, 2003; Beinhocker, 2006; Sawyer, 2005; Elder-Vass, 2010, 26–53; Padgett and Powell, 2012). We can recognize complex systems by their shared, signature characteristics, such as power laws, volatility, and self-similarity in the form of fractals (Beinhocker, 2008, 163).

1.8.2 Complexity and Emergence

Complexity theory is the opposite of reductionism. Whereas reductionist scientists wend their way down the hierarchy of life—from large-scale structures such as nation-states to the smallest entities such as atoms and molecules—complexity theorists follow the activity of entities as they move upward as well, from one layer of a system's hierarchy to the next, all the while gaining greater fitness (Fromm, 2004; Kauffman, 2019, 5). The process by which entities ascend higher is called emergence, and it is this process, together with adaptation, that constitutes the mechanisms driving social change[19] (Kauffman, 1993; Holland, 1996). Because standards govern interactions and provide a reference map for adaptation, they are essential to this process.

1.8.3 The Adaptability of Complex Systems

The indeterminateness and flexibility associated with complexity allow complex adaptive systems to evolve and adapt over time. In fact, as Beinhocker claims, complex adaptive systems are, by their very nature, evolutionary systems, which are ideally suited to repeated learning (Beinhocker, 2006). Learning takes place when actors at the micro level strive to enhance their fitness level with respect to the context (the fitness landscape) in which they are situated[20] (Monge and Contractor, 2003). In so doing, they change the fitness levels of other actors as well as the fitness landscape of the system itself—that is, the macro criteria by which actors in that system are evaluated (Kauffman, 1995; Beinhocker, 2006). It is in this way that actors and systems co-evolve.

1.8.4 The Rules of Emergence

Viewed in terms of complex adaptive systems, *standards are the rules of emergence.* They define fitness levels and fitness landscapes by explicating the criteria for success in any given context. Like all norms and institutions, standards emerge and coevolve through the interplay of social interactions, social institutions, and social norms, be they cultural, political, or economic. Embedded in language, artifacts, scripts, and repertoires, standards serve as *interfaces* between actors at all layers of complex adaptive systems. Facilitating interconnection and interaction, they foster the generation of emergent properties and the evolutionary adaptation of the system itself (Kontopoulos, 1993; Sawyer, 2005; Beinhocker, 2006; Holland, 2014).

1.8.5   *What Makes Standards Mechanisms Unique?*

Hence, standards are mechanisms much like enzymes in the molecular world. What distinguishes standard mechanisms from the laws studied by natural scientists is the extent of their indeterminateness. While human agents may skirt existing standards or impose new ones, depending on the circumstances, nature is more immutable. In the absence of mutations and evolutionary adaptations, natural laws lead to predictable outcomes. In contrast, social standards cannot generate high levels of predictability, but their flexibility allows them to be updated incrementally, without major disruptions, as social conditions change (Kauffman, 1995; Bak, 1996). Keep in mind, however, that the abrupt alteration of a critical mass of standards is itself a change mechanism. When many standards are displaced simultaneously, they give rise to phase transitions that open up new opportunities and possibilities for social action together with new standard mechanisms (Bak, 1996, 134–137).

1.9   Network Architecture

Standardized components do not exist in isolation. They are connected by protocols to other diverse components, forming clusters and networks[21] that serve as layered platforms (Gawer, 2009). Together, these networks comprise a hierarchical structure such that each layer is made up of its own unique components, which are governed by their own, specific protocols. This structure and the rules that govern how components function within the network constitute the network's architecture. Although some networks are more pronounced and concrete than others, all have an architecture, which serves to structure and constrain the ways in which interaction takes place (van, Schewick, 2012; Garcia, 2005).

Hence, a network's architecture can have a considerable impact not only on its users but also on the social system in which it is embedded. Characterizing the effects on economic actors and a nation's economy, for example, van Schewick (2012) notes:

> Different architectures may impose different constraints, which may result in different decisions by economic actors, which in turn may result in different firm and market structures and different levels of economic activity. And by changing existing architectures or creating new ones, economic actors can change the constraints that architectures impose. (xvii)

1.10   Standards Beget Standards

Not surprisingly, standards have proliferated and gained importance as society has become more complex. As Emile Durkheim noted, increased specialization and a deeper division of labor generate the need for greater integration and control (Durkheim, 1984). Standards provide a useful solution to meet this need (Beniger, 1989). However, while successful in the short term, employing standards to reduce complexity in the long run may be self-defeating. As standards reduce complexity, actors typically extend and deepen their interactions, bringing new actors and enterprises into the fold. To integrate and support these newcomers, new standards will be required. As well, new ways of generating standards might be called for. In this way, standards not only beget standards; as they winnow away outdated standards, they ensure that standards coevolve. It is by doing so that standards generate emergence and greater fitness higher up the fitness landscape.

1.11 The Framework for Analysis

Analyzing standards and their impact on complex systems requires a generic, interdisciplinary perspective. In all realms, standards are analogous in how they govern interactions. Likewise, notwithstanding standards' specific differences, they all come about in similar ways. Given these commonalities, we can analyze standards from atop the ladder of abstraction, employing the same analytical framework to study their roles and impacts in diverse cases.

This book seeks to determine how standards, by generating emergence, constitute the missing link between cause and effect in complex systems so as to generate evolutionary changes. Because most social science literature is focused on stasis, it is not particularly helpful in this regard. Economists, for example, strive for equilibrium; sociologists seek integration; and anthropologists emphasize the importance of norms and common values in promoting stability. To understand the dynamics of change and the link between cause and effect we must look more holistically through the lens of evolution, and the longue durée,[22] in the context of complexity.

To pursue this approach, I start with a conceptual platform[23], much like the standards platforms described above. At its core is the nexus between standards, complexity, emergence, and evolutionary change. In this model, standards are the mechanism of change; complexity is the context in which change occurs; emergence is the process by which change takes place; and evolution is the outcome of change. To ensure a holistic perspective, I link the concepts at the core of the platform to the peripheral "adjacent possible",[24] where a range of disciplinary perspectives are located. These weaker links allow me to draw from a wide range of diverse disciplinary perspectives depending on the issue at hand.

In the following chapters, I lay out empirical cases, each of which identifies and describes the often-neglected power of standards to orchestrate interactions and trigger changes. For my evidence, I adopt a historical, sociological approach, relaying stories about standards and their role in generating emergence in complex environments (Braudel, 1980; Mahoney and Rueschemeyer, 2003; Kim, 2016). Focusing on diverse historical episodes, I compare standards and standard-setting processes so as to identify their commonalities and differences as they apply to emergent outcomes (Lachman, 2013). In so doing, I relate how standards both generate and guide emergence, leading to greater complexity and higher performance, with evolution as the result (Kauffman, 1995, 2003).

Each chapter in the book provides a story on its own and can be read independently of the others. However, each chapter builds upon the preceding ones such that together they provide a complete picture of the nature of complexity and complex systems. Much as complexity theory portends, the "whole is greater than the sum of the parts".

Notes

1 Whereas Tilly speaks of transactions, I prefer the term interactions, which has less of a utilitarian connotation.

2 Note that the narratives described herein are not limited to a focus on societal interactions. Referencing biological entities, for example, Levinson notes: In the macroscopic realm where interactions take place, every imaginable event that can be observed—at least in theory—involves interactions between two or more objects (Levinson, 2004, 44). Likewise, as Mayfield attests: "In the world of physics, meaning emerges from interactions between system states. If there are no interactions, there is no meaning. For meaning to be present, particular states of one system must have effects on another system" (Mayfield, 2013, 58).

3 Throughout this volume, the term social mechanism refers to the link between cause and effect.

4 See, for example, *Standards, Building Blocks for the Future* by US Congress, Office of Technology Assessment (1992), Washington, DC, US Government Printing Office (Garcia, 2002, 2005).

5 The concept of phase transitions is derived from physics and biology. It refers to a shift in matter from one state to another, as, for example, in the freezing of water into ice. We can also speak of phase transitions in terms of social, economic, and cultural events. Consider, for example, the fall of Rome and its replacement by the barbarian kingdoms of the Middle Ages. For such an example, see Chapter 4. For a scientific explanation, see Bak (1996).

6 Externalities refer to the costs and benefits associated with an activity or technology due to external factors. For example, the interdependencies of networking components give rise to both positive and negative externalities. Thus, up until a certain point, adding another participant to a network will likely enhance the value of the network for existing participants. Similarly, new network applications are likely to increase the value of existing applications as well as the value of the network to users. On the other hand, to the extent that they cause network congestion, additional applications and users can give rise to negative externalities. These interdependencies and externalities explain why networks, once they gain momentum, assume path-dependent trajectories (Arthur 1989, 161–131).

7 The Scopes trial took place in the summer of 1925 in the small town of Dayton, Tennessee. On trial was a local teacher, John Scopes, who was charged with teaching evolution to his class. What made the trial so significant was the two eminent lawyers who argued the case. Williams Jennings Brian, an evangelist and former presidential candidate, argued for the prosecution, while Clarence Darrow, an atheist and famous defense lawyer, argued, together with the ACLU, on behalf of the defendant. See Edward J. Larson (2022).

8 The term "social facts" was coined by sociologist Emile Durkheim in reference to the social structures within society that serve as the constraints and affordances that determine individual and group agency.

9 As defined by Bunge, "A *mechanism* is a set of processes in a system, such that they bring about or prevent some change—either the emergence of a property or another process—in the system as a whole" (Mario Bunge, 2003).

10 Reductionism is a methodological approach that seeks to explain complex outcomes by generalizing based on the nature of their most diminutive, individual parts.

11 In its purest form, sociobiologists believe that human nature can be explained by the evolution of individual genes within the DNA. As Hoynegen-Huene describes, sociobiology is based on the assumption that "living beings are made up of the same sort of atoms and molecules as the physicists and chemists know them, and nothing else, [so that] having a comprehensive knowledge of the laws of physics and chemistry would in principle be sufficient to redefine all the properties of [an] organism in atomic and molecular terms, and to derive its behavior—specifically, all the laws by which it behaves (Hoynigen-Huene, 1989, 43).

12 According to Walter Buckley, for example, "Objections that many beside myself have too much sociological theory today are that it is faddish; it is often built on abstruse notions that defy empirically meaningful definition; it is overly impressed with exquisitely contorted writing—especially from people in prestigious universities or in exotic conclaves in Europe" (Buckley, 1998).

13 Methodological individualism is an analytical approach that centers on the "autonomous" individual, rather than the group. Typically, scholars in this category have focused on decision criteria that focus on efficiency and effectiveness and maximizing the gains to the individuals involved.

14 One way of reconciling methodological individualism with a more holistic approach is to follow Paggett's and Powell's conceptualization of the social order. As the authors describe it: "*In the short run actors create relations; in the long run, relations create actors*". As they say, "The difference between methodological individualism and social constructivism is not a matter of religion; it is a matter of time scale" (Padgett and Powell, 2012, p. 27). Because this volume seeks to account for the factors driving evolution, its time frame consists, for the most part, of the *longue durée* as conceived by Fernand Braudel and the Annales School (Lee, 2012).

15 Writing *Suicide (1897)* Emile Durkheim developed the concept of social facts to explain the differential rate at which Catholics and Protestants committed suicide. For him, social facts included such things as norms, values, rules, and institutions, which constrained individuals in their behavior. He contended that social facts should serve as the unit of analysis in sociology. In my analyses, I treat standards as serving as social facts.

16 Durkheim conceived of social facts as social structures. Simmel conceived of structures as the crystallization of interactions. In his characterization, John Levi Martin (2009) includes the effects of social structures. As he notes, "… social interactions, when repeated, display formal characteristics; and this

form can then take on a life of its own, ultimately leading to institutions that we (as actors) can treat as given and exogenous to social action for our own purposes, though at any moment (or at least at some moments) these institutions may crumble to the ground if not rejuvenated with compatible action" (Martin, 2009, 18).

17 John Holland and Murry Gell-Mann coined the term "complex adaptive systems". The term refers to the ideas and propositions developed at the Santa Fe Institute, as well as the work of scholars who subsequently emerged to pursue the field of complexity. As defined by SFI, complex systems are those in which the "whole is greater than the sum of the parts", and in which outcomes cannot be traced back to the parts (see Waldrop, 1995, for a history of the Santa Fe Institute and its work relating to complex adaptive systems).

18 The role of standards in relationship to these attributes is laid out in Chapter 12.

19 Social change can also come about in a phase transition, as the result of a system's failure to adapt to its changing environment.

20 Fitness levels measure the degree to which an actor's behaviors match the needs of the environment. Fitness landscapes constitute the criteria of success (Kauffman, 2003).

21 Interactions are ubiquitous throughout the universe. When these interactions are linked together, they give rise to networks. It is common standards that link networks together and determine their capabilities and architecture. Because networks give rise to externalities, they can exercise considerable power. For a discussion, see David Singh Grewal (2009), *Network Power*.

22 The term is one employed by Fernand Braudel to differentiate long-term structural changes from short-term events. See Braudel (2009). He argued that to adopt a long-term perspective, one needs to employ an interdisciplinary approach. I contend that a focus on standards facilitates such an inquiry.

23 Much like today's digital platforms, this platform consists of a tightly knit, stable conceptual core—standards, emergence, complex outcomes—that is linked to diverse, flexible external inputs drawn from a larger body of knowledge. It is in this way that the conceptual platform provides a means of analyzing a broad variety of scenarios, as is employed throughout this volume.

24 The "adjacent possible" is a concept developed by Stewart Kauffman to describe an area in close proximity to another that houses complementary information and ideas that are useful to a core area. For another discussion that relates the adjacent possible to innovation (Kauffman, 1993; see also Johnson, 2010).

References

Anderson, Benedict (2016) *Imagined Communities; Reflections on the Origin and Spread of Nationalism*, New York, NY: Verso.

Arthur, Brian (1989) ""Competing Technologies, Increasing Returns, and Lock-in by Historical Events," *Economic Journal* 99, 116–131.

Arthur, Brian (2009) *The Nature of Technology: What It Is, and How It Evolves*, New York, NY: Free Press.

Bak, Per (1996) *How Nature Works: The Science of Self-Organized Criticality*, New York, NY: Oxford University Press.

Beinhocker, Eric D. (2006) *The Origin of Wealth: The Radical Remaking of Economics and What it Means for Business and Society*, Cambridge, MA: Harvard Business School.

Beniger, James (1989) *The Control Revolution: Technological and Economic Origins of the Information Society*, Cambridge, MA: Harvard University Press.

Biddle, B.J. (1986) "Recent Developments in Role Theory," *Annual Review of Sociology* 12, 67–92.

Braudel, Fernand (1980) *On History*, Chicago, IL: University of Chicago Press.

Braudel, Fernand, and Immanuel Wallerstein (2009) "*History and the Social Sciences: the longue durée,*" *Review (Fernand Braudel Center)*, 32(2), 171–203.

Brennan, Geoffrey, Lina Eriksson, Robert E. Goodin, and Nicholas Southwood (2013) *Explaining Norms*, New York, NY: Oxford University Press.

Buchanan, Mark (2007) *The Social Atom, Why the Rich Get Richer, Cheaters Get Caught, and Your Neighbor Usually Looks Like You*, New York, NY: Bloomsberry USA.

Buckley, Walter (1998) *Society—A Complex Adaptive System; Essays in Social Theory*, Amsterdam: Gordon and Breach Publishers.

Bunge, Mario (2003) *Emergence and Convergence: Qualitative Novelty and the Unity of Knowledge*, Toronto, Canada: University of Toronto Press.

Grewal, David Singh (2009) *Network Power: The Social Dynamics of Globalization*, New Haven, CT: Yale University Press.

Demeulenaere, Pierre, ed. (2011a) *Analytical Sociology and Social Mechanisms*, New York, NY: Cambridge University Press.

Demeulenaere, Pierre (2011b) "Causal Regularities, Action, and Explanation," in Demeulenaere, Pierre, ed. *Analytical Sociology and Social Mechanisms*, New York, NY: Cambridge University Press.

Durkheim, Emile (1984) *The Division of Labor in Society*, New York, NY: Palgrave, Macmillan.

Easley, Kleinberg (2010) *Networks, Crowds, and Markets*, New York, NY: Cambridge University Press.

Elder-Vass, Dave, ed. (2010) *The Causal Power of Social Structures—Emergence, Structure and Agency*, New York, NY: Cambridge University Press.

Elster, Jon (2015) *Explaining Behavior: More Nuts and Bolts for Social Sciences*, New York, NY: Cambridge University Press.

Fromm, Jochen (2004) *The Emergence of Complexity*, Kassel, Germany: Kassel University Press.

Garcia, D. Linda (2002) "The Architecture of Global Networking Technologies," in Sassen, Saskia, ed. *Global Networks: Linked Cities*, New York, NY: Routledge.

Garcia, D. Linda (2005) "Bringing the Public Interest into Standard Setting: The View from a Complexity Perspective." Report prepared with grant from Sun Microsystems.

Garcia, D. Linda (2016) "The Evolution of the Internet: A Socioeconomic Account," in Johannes, Bauer, and Michael Latzer, eds. *Handbook on the Economics of the Internet*, Cheltenham: Edward Elgar.

Gawer, Annabelle, ed. (2009) *Platforms, Markets and Innovation*, Cheltenham: Edward Elgar.

Granovetter, M. (1985) "Economic Action and Social Structure: The Problem of Embeddedness," *The American Journal of Sociology* 91 (3), 481–510.

Hankins, Barry (2010) *Jesus and Gin: Evangelicals, The Roaring Twenties and Today's Cultural Wars*, New York, NY: Macmillan.

Hedstrom, Peter (2005) *Dissecting the Social, on the Principles of Analytical Sociology*, New York, NY: Cambridge University Press.

Holland, John H. (2014) *Signals and Boundaries: Building Blocks for Complex Adaptive Systems*, Cambridge, MA: MIT Press.

Hoyningen-Huene, Paul, and Franz M. Wuketits, eds. (1989) *Reductionism and Systems Theory in the Life Sciences: Some Problems and Perspectives*, New York, NY: Kluwer Academic Publishers/Springer.

Isaiah 18 (1611) *Holy Bible*. King James Version.

Johnson, Steven (2010) *Where Do Good Ideas Come From: The Natural History of Innovation*, New York, NY: Bloomberg.

Katz, Daniel and Robert L. Kahn (1978) *The Social Psychology of Organizations*, New York, NY: Wiley.

Kauffmann, Stuart (1993) *The Origins of Order: Self-Organization and Selection in Evolution*, New York, NY: Oxford University Press.

Kauffman, Stuart (1995) *At Home in the Universe: The Search for Laws of Self- Organization and Complexity*, New York, NY: Oxford University Press.

Kauffman, Stuart (2008) *Reinventing the Sacred, A New View of Science, Reason and Religion*, Philadelphia, PA: *Basic Books*.

Kauffman, Stuart (2019) *A World Beyond Physics: The Emergence and Evolution of Life*, New York, NY: Oxford University Press.

Kim, Jeong-Hee (2016) *Understanding Narrative Inquiry*, Thousand Oaks, CA: Sage Publications, Inc.

Kirschner, Mark W. and John C. Gerhart (2005) *The Plausibility of Life: Resolving Darwin's Dilemma*, New Haven, CT: Yale University Press.

Kontopoulos, K. (1993) *The Logic of Social Structure*, New York, NY: Cambridge University Press.

Lachman, Richard (2013) *What Is Historical Sociology?* Malden, MA: Polity Press.

Larson, Edward J. (2022) *Summer of the Gods, the Scope Trial and the Continuing Debate over Science and Religion*, New York, NY: Basic Books.

Lee, Richard E. (2012) *The Longue Durée and World System Analysis*, New York, NY: State University of New York Press.

Levinson, Gene (2004) *Rethinking Evolution, the Revolution That's Hiding in Plain Sight*, Hackensack, NJ: World Scientific Publishing.

Mahoney, James, and Rueschemeyer Dietrich (2003) *Comparative Historical Analysis in the Social Science*, New York, NY: Cambridge University Press.

Malone, Kemp (December 1932) "Observations on the Word 'Standard'," *American Speech* 17 (4).

Martin, John Levi (2009) *Social Structures*, Princeton, NJ: Princeton University Press.

Mayfield, John (2013) *The Evolution of Complexity*, New York, NY: Columbia University Press.

McNichol, Tom (2011) *AC/DC: The Savage Tale of the First Standards War*, New York, NY: John Wiley & Sons.

Monge, Peter, and Noshir Contractor (2003) *Theories of Communication Networks*, New York, NY: Oxford University Press.

Padgett, John F., and Walter W. Powell, eds. (2012) *The Emergence of Organizations and Markets*, Princeton, NJ: Princeton University Press.

Parker, Geoffrey G., Marshall W. Van Alstyne, and Sangeet Paul Choudhry (2016) *Platform Revolution: How Networked Markets Are Transforming the Economy—and How to Make Them Work for You*, New York, NY: W.W. Norton and Company.

Rolfs, J. (2001) *Bandwagon Effects in High Technology Industries*, Cambridge, MA: MIT Press.

Rosenau, James, and Mary Durfee (1999) *Thinking Theory Thoroughly: Coherent Approaches to an Incoherent World*, 2nd edition, New York, NY: Routledge.

Ryan, Johnny (2010) *A History of the Internet and the Digital Future*, New York, NY: Reaktion Books Ltd.

Sawyer, R. Keith (2005) *Social Emergence; Societies as Complex Systems*, New York, NY: Cambridge University Press.

Shapiro, Carl, and Hal R. Varian (1998) *Information Rules: A Strategic Guide to the Network Economy*, Brighton, MA: Harvard Business School Press.

Sheldrake, Merlin (2020) *Entangled Life: How Fungi Make Our Worlds, Change Our Minds, and Shape Our Futures*, New York, NY: Random House.

Sperber, Dan (2011) "A Naturalistic Ontology for Mechanistic Explanations in the Social Sciences," in Demeulenaere, Pierre, ed., *Analytical Sociology and Social Mechanisms*, New York, NY: Cambridge University Press.

Stokes, David (2011) "Jesus and Gin: Evolutionism, The Roaring Twenties, and Today's Cultural Wars." https://www.seattlepi.com/lifestyle/blogcritics/article/Book-Review-Jesus-and-Gin-Evangelicalism-the-890846.php

Tilly, Charles (2016) *Identities, Boundaries and Social Ties*, New York, NY: Routledge.

Thomson, Irene Taviss (2010) *Cultural Wars and Enduring American Dilemmas*, Ann Arbor, MI: University of Michigan Press.

Turner, Jonathan H., and Richard S. Machalek (2018) *The New Evolutionary Sociology: Recent and Revitalized Theoretical Approaches*, New York, NY: Routledge.

US Congress, Office of Technology Assessment (1992), *Standards: Building Blocks for the Future*, Washington, DC: US Government Printing Office.

US Congress, Office of Technology Assessment (1994) *Electronic Enterprises: Looking to the Future*, Washington, DC: US Government Printing Office.

van Schewick, Barbara (2012) *Internet Architecture and Innovation*, Cambridge, MA: MIT Press.

Waldrop, Mitchell M. (1995) *Complexity: The Emerging Science at the Edge of Order and Chaos*, New York, NY: Open Road.

Weber, Max (2001 [1930]) *The Protest Ethic and the Spirit of Capitalism*, New York, NY: Routledge.

West, Geoffrey (2018) *Scale the Universal Laws of Life, Growth, and Death in Organisms, Cities and Companies*, New York, NY: Penguin Press.

Williamson, Oliver E. (1985) *The Economic Institutions of Capitalism*, New York, NY: Macmillan.

Winner, Langdon (2020) *The Whale and the Reactor, a Search for Limits in the Era of High Technology*, London: University of Chicago Press.

Wu, Tim (2010) *The Master Switch: The Rise and Fall of Information Empires*, New York, NY: Alfred A. Knopf.

Wuketits, Franz, M. 1989 (1989) Organisms, Vital Forces, and Machines: Classical: Controversies and the Contemporary Discussion 'Reductionism vs. Holism,' in Hoyningen-Huene, Paul, and Franz M. Wuketits, eds. *Reductionism and Systems Theory in the Life Sciences, Some Problems and Perspectives*, New York, NY: Kluwer Academic Publishers/Springer.

2

ASCENT UP THE FITNESS LANDSCAPE

How the American West Was Won

2.1 Introduction

Four years ago, my husband and I moved from the hectic capital city of Washington, DC to the neighborly town of La Grande, Oregon. Located in the northeastern part of Oregon, in a valley bounded by four mountain ranges, La Grande is a small town, much like Richard Scarry's "Busytown," described in the Preface. Its claim to fame was its role in the migration West. La Grande was one of the first stops the pioneers made in that magical Oregon of their destination and dreams. So happy, after hundreds of miles of sagebrush desert, to have arrived in such a fertile valley, many pioneers decided to stay in La Grande, rather than struggle on. One can follow their trek by the ruts left in the trail—preindustrial traffic signs, so to speak—which are still visible today. Nor can one ignore the graves that dotted the 2000-mile trail—*signals* to the pioneers of the dangers that lay ahead (Dary, 2007). Estimates are that between 20,000 and 30,000 emigrants died along the way (Oregon Interpretive Center, visit, 2018; Ontko, 1997, 13). Living with the ghosts of the pioneers, I have researched their histories and diaries, as well as visited their sights to better engage with their history (Kesselman,1974; Applegate and O'Donnell,1994; Beckham, 2006).

One of the major accounts of the American movement west is President Theodore Roosevelt's four-volume epic history, *The Winning of the West 1889–1896* (Roosevelt, 1889). Encouraging Americans to fulfill their "manifest destiny," Roosevelt elaborately detailed the drama, determination, and daring adventures of those pioneers, cowboys, and cavalry who led America's westward expansion. Notwithstanding Roosevelt's evocative portrayal, I found something missing. How, I wondered, did all these actors come together and coordinate their activities? How did they develop relationships with a diverse group of Europeans as well as a broad array of Native Americans with whom they had such cultural differences?

Far less gripping and exhilarating than the early dramatic tales of the West—although by no means less significant—is the untold story of standards, and how, operating under the radar, they provided a platform for this great westward migration. Recall that I have defined standards as interfaces that govern all interactions, whether between people, machines, or people and machines. With this idea in mind, picture the vast stretch of territory making up the North American

DOI: 10.4324/9781032721125-4

continent. Imagine, then, the boundless number of standardized interactions required to pave the way west. As we shall see, while often *ad hoc* in nature and negotiated *en route*, standards emerged as the platform upon which, and according to which, travelers journeyed; battles were lost and won; trade was established; and a frontier culture was born.

2.2 Complexity, Standards, and Emergent Behavior

Much like Roosevelt's telling, many writings about the migration westward are based on chronological accounts that unearthed the role of critical actors as well as the impact of major events. However, I am more inclined to view this epic relocation from a bird's-eye view, examining the "whole" as opposed to "the individual parts." To this end, I characterize this movement in terms of complexity, portraying it as an evolutionary process linked by standardized interactions, and fueled by emergent behavior, which led over time to a coherent system exhibiting greater fitness and higher levels of performance (Kauffman, 2008; Beinhocker, 2007; Johnson, 2010).

2.2.1 Augmented Fitness Levels

In moving west, the pioneers gradually worked their way up the *fitness landscape*.[1] The *fitness landscape* refers to the context in which an actor must adapt to survive. Embodying the local culture as well as the structural characteristics of a given context, the *fitness landscape* defines the criteria for success in any given situation. The measure of an actor's ability to match its resources to a particular context constitutes the actor's *fitness level*. Because the landscape and environment through which the pioneers traveled varied from place to place, they had to accumulate greater skills, resources, and fitness as they moved onward (Ernst and Kim, 2001; Vandenbroucke, 2008; Pitzer et al., 2011). Adopting new standards suited to each new challenge was critical to this adaptive process. In terms of complexity theory, we might say that the Western migration followed the path of the "adjacent possible," a term coined by complexity biologist Stuart Kauffman[2] (Kauffman, 2008, 80–81; Johnson, 2002, 14–25). That is to say, with each step forward into virgin territory, new worlds were discovered requiring different skills, roles, protocols, and technologies (Billington and Ridge, 2001, 45). By embracing new standards, pioneers established the wherewithal to operate successfully in one locale before advancing to the next new frontier (Billington and Ridge, 2001, 2; Dary, 2007). Emphasizing the importance of the *process of migration*, Hine et al. (2000) note:

> Whatever its boundaries, in American history, the West is not only a modern region somewhere beyond the Mississippi but also the process of getting there. That may make the western story more complicated, but it also makes is more interesting, more relevant. The history of the frontier is a unifying American theme, for every part of the country was once a frontier, every region was once a West. (11)

2.2.2 Reaching a Threshold

These spurts in the migratory process are in keeping with Mark Granovetter's notion of "*thresholds*," which he employs to explain collective behavior such as mass emigrations (Granovetter, 1987, 1420–1443). In contrast to many accounts of collective behavior, Granovetter does not assume that all participants in a joint venture share the same norms and interests. Instead, he posits that, in a diverse interacting group, people have different thresholds according to which they are willing to join a movement or not. Granovetter defines these thresholds as the points at which the benefits of joining outweigh the costs. Although thresholds differ among diverse participants depending on their stakes in a particular undertaking, their thresholds typically fall as the number of participants increases, giving rise to what social scientists refer to as "cumulative causation." This pattern occurs because more participants signal less risk and greater gains (Granovetter, 1987). When this happens, the process becomes self-reinforcing, and fitness levels rise.

2.2.3   Emergence

Once the pioneers were well prepared and resolved in their decision to go west, much like a spark—set them on their way. That spark ignited the process of emergence. The concept of emergence refers to self-organization that is generated by bottom-up interactions (Holland, 1999; Kauffman, 2008; Johnson, 2010, 13). As Steven Johnson (2002) describes in his book *Emergence, the Connected Lives of Ants, Brains, Cities and Software,* emergent systems "solve problems by drawing on masses of relatively stupid elements rather than a single intelligent 'executive' branch." As John Holland points out, this process occurs when individual elements collectively follow a set of simple rules. Comparing the phenomenon of emergence to a board game, Holland (1999) notes:

> Agreement on a few rules gives rise to extraordinarily complex games. Chess is defined by fewer than two dozen rules, but humankind continues to find new possibilities in the game after hundreds of years of intensive study. …The hallmark is the sense of much coming from little. (1)

2.2.4   The Role of Standards

And so it was with the "Pioneers" journey West. Without any overall coordination, and sometimes only the feedback of word of mouth, one step in the migration led to yet another, and then again another. How did this happen? Standards, developed *en route,* provided the rules! Remember that emergent processes, such as those described above, are propelled by interactions at multiple levels. As importantly, interactions require standard interfaces, in the form of codes, signs, signals, norms, memes, etc., to take place (Holland, 2014). It is in this way that standards, when networked together, generate a platform for emergent processes, as well as a channel for evolutionary adaptation to take place (Holland, 2014). Hence, the mastery of the West could not have occurred without a countless number of new standards that addressed new conditions and realities. In the description that follows, I explore how, and under what circumstances, these standards came about.

2.3   A Visionary Meme

In the movement west, early emigrants paved the way for an ever-growing number of pioneers. Over a period of years, a critical mass was reached, engendering a phase transition—that is to say, a radical break with the past and a new order for the future (Billington and Ridge, 2001, 3; Vandenbroucke, 2008). At this point, we might say "the American West" was born.

However, long before it was a reality, the "West" was but a visionary idea (Smith, 1990; Hine et al., 2000; De Voto, 2019, 23). Jefferson embraced this vision, and, on becoming president, he charged Meriwether Lewis and William Clark to undertake an exploratory expedition into the unknown miles of rivers, deserts, forests, and mountains that gave way to the Pacific Coast (Beck and Haase, 1989; Billington and Ridge, 2001; Dary, 2007, 26). As Henry Nash Smith explains, the Lewis and Clark expedition gave rise to a persistent, standardized *meme* characterizing the West as the promised land. As he recounts:

> The importance of the Lewis and Clark expedition lay on the level of imagination: it was drama, it was the enactment of a myth that embodied the idea and established the image of a highway across the continent so firmly in the minds of Americans that repeated failures could not shake it.
>
> *(Smith, 1990, 17)*

The proliferation of this all-encompassing meme meant that, much as Granovetter might have surmised, diverse groups of people could each project their own, particular version of paradise onto their vision of the West. Hence, notwithstanding their differences, they had a solid basis for cooperation. But equally important, given the difficulty of the trip and the uncertainties it entailed, each needed to share their individual knowledge so that the whole group would be prepared. For it is likely that each individual that journeyed west embodied some tacit knowledge of diverse aspects of trailblazing. Their knowledge became explicit as they acted upon it and shared it in conjunction with others. Over time, this explicit knowledge was embedded in standard ways of doing things (Polanyi, 1962; Brown and Duguid, 2017). By adopting these standards, newcomers could build on the experiences and the cumulative knowledge of those who had come before (see Holland, 2014, 57–58). As described by Billington and Ridge (2001):

> The story of westward expansion was one of the continuous reconstruction of society, of repeated 'beginning over again' in the West with the same results on every frontier, although with differences due to time, place, and the manner of men and women who peopled the area. (2)

2.4   Dividing the Pie among the Players

While standards can generate cooperation, they can also create dissent. The process of standard setting is not neutral. Some will benefit, while others will lose (Heckathorn,1996; Kolloch, 1997; Gulati and Singh, 1998; Boone and Macy, 1999). Thus, while it was one thing for the pioneers to have adhered to a common meme in their efforts to move west, it was quite another for them to divide up the western territory in accordance with their diverse aspirations and needs. Given competing interests, it is not surprising that the early history of the Far West was dominated by conflicts over boundaries and the process of establishing and enforcing them *via* standards. By the time the pioneers arrived in the west, there were many more interested players with widely diverging interests.

2.4.1   *The Native Americans*

When the first explorers arrived in the Northwest, they encountered a wide array of Native Americans, whose history in the region was well-established, dating back 10,000 years. Estimates are that about 180,000 natives lived in this region. Distributed among 2000 diverse cultures and approximately 125 tribes, they had relatively limited contact with one another, although they shared, for the most part, a common way of communal life (Hine et al., 2000; Billington and Ridge, 2001, 50–60; Dary, 2007, 16–17).

The arrival of the horse *via* the Spanish conquistadores greatly altered their fitness landscape (Beck and Haase, 1989). Prior to the horse, Native Americans were typically confined to the small areas where they lived off the land and local game. However, when propelled by the horse, these Indians could travel vast distances, encountering other tribes with whom they might trade and share stories and information (Ontko, 1997, 24–25; Hine et al., 2000; Dary, 2007, 15). Moreover, instead of stalking small game, they could—from atop the horse and at top speed— hunt the buffalo, which were not only prolific but also provided for all the Native Americans' needs (Beck and Haase, 1989, 9; Ontko, 1997, 24). Loosely allied together into larger nations, such as the Shoshoni and the Nez Perce, individual tribes remained in contact with one another, joining together to celebrate major festivities or to fight common enemies.

Up until the 19th century, the many geomorphic boundaries that inhibited travel across the American continent reinforced the security and fitness of the western tribes. Outsiders were few and far between. Travel beyond the Mississippi River was treacherous. The terrain was far different from anything Easterners had seen (Beck and Haase, 1989) Travelers had first to negotiate their way across the Great American Desert that, although relatively flat, was subject to hot, dry winds, sandstorms, quicksand, and waves of locusts. The mountain ranges were even more foreboding, as were the deserts further on. As one traveler commented on seeing the Rockies:

Nature has fixed limits for our nation, she kindly introduced as our western barriers, mountain almost inaccessible, whose base she has skirted with irreclaimable deserts of sand.

(Beck and Haase, 1989, 6)

Hence, by the 1840s, when the Americans first arrived in large numbers, the northwestern tribes were relatively fit (Beck and Haase, 1989). Describing the powerful Shoshoni, Ontko (1997) notes:

The Shoshoni had now completed the pleasant transition from passive, starving, dirty proletarians to plump, clean aggressive capitalists with time to explore or go fishing just for the fun of it. Their hunting parties probed the Great Plains in pursuit of buffalo and the Pacific Coast in quest of trade articles with no fear of retaliation. To stave off boredom they fought among themselves. (32–33)

But the fitness landscape was about to change! The British, French, Russians, and Spanish all had claims and settlements in the New World (Billington and Ridge, 2001). Their fierce competition was governed by standards of diplomacy and warfare. However, as in most complex situations, outcomes were determined not solely by local events but also, and as importantly, by circumstances well beyond their bounds and control (Billington and Ridge, 2001, 31–44). For the Native Americans, the entanglement with the Europeans was catastrophic, leading to a loss of their land as well as their way of life (Ontko, 1997) For the Europeans, the conflicts in Europe, together with the American Revolutionary War, contributed, in large measure, to the redistribution of much of the New World territory to the newly established United States of America (Beck and Haase, 1989, 15–20; Ontko, 1997; De Voto, 2019, 23–24).

2.4.2 The Mountain Men

When standards are inconsistent, translators are required. One need only think of a modem, which modulates and demodulates signals so they can travel from analog to digital devices and *vice versa*. In a similar fashion, the American pioneers, traveling west, needed translators to help them bridge the gap between European standards and those of the frontier.

Enter the mountain men (De Voto, 2019, 85–103). Describing the importance of their role in the migration west, Billington and Ridge (2001) note:

> Those unofficial explorers spied out the secrets of all the West, plotted the course of its rivers, discovered the passes through its mountains, and prepared the way for settlers by breaking down the Indian self-sufficiency. (93)

The mountain men were among the earliest European explorers in the New World. Like Davy Crockett, they preferred to live on the outskirts of society, delighting in the freedom, authenticity, and simplicity of the wilderness (Smith, 1990). As new settlements encroached upon them, they moved further and further west. These mountain men learned many of their wilderness skills from Native Americans, often cohabiting with them, and adopting their styles of dress, learning their languages, marrying their women, and even becoming members of their tribes (Smith, 1990; Hine et al., 2000). In fact, a common saying was that "while an Indian could never become a white man, a white man could easily become an Indian" (De Voto, 2019, 99).

To make a living, mountain men trapped and traded beaver and otter pelts, a relatively lucrative endeavor, as these furs were not only plentiful in the West but also highly desired by consumers in Europe and the Far East. Organizing these trappers were entrepreneurial traders, who set up trading posts, called factories, where trappers converged to trade. Most elaborate in this regard were the annual rendezvous of traders and trappers who, much as in the Medieval Fairs in Europe, assembled to trade not only their pelts and furs but also their stories and discoveries. These rendezvous were especially effective because they embraced both the Indian's and the Mountain Men's ways of life (Billington and Ridge, 2001, 99). In the process, they produced common norms and practices that governed their actions and provided relative peace and stability (Becke and Haase, 1989, 26). As importantly, the rendezvous contributed to what might be described as a platform and burgeoning infrastructure over which settlers later traveled (Billington and Ridge, 2001).

In time, however, overzealous trapping led to the decline of the fur trade. Forced to find new means of livelihood, many mountain men became guides and leaders of the Pioneer Conestoga Wagons traveling west (Billington and Ridge, 2001; Dary, 2007, 113–114). Also preparing the way west were the missionaries and the military men, many of whom had helped to inspire and escort the Pioneers.

2.4.3 The Missionaries

Hoping to convert Native Americans to Christianity and the standards that it espoused, the missionaries established their headquarters along the Oregon and Santa Fe Trails (Billington and Ridge, 2001, 158–160; Dary, 2007, 67–68). Encouraging settlers to migrate, their correspondence and reports back to the East characterized the Oregon Territory as an agricultural paradise. Although their appeals to the settlers were successful, the missionaries were far less effective in converting Native Americans to Christianity. Unlike the traders and trappers who had compromised with and adapted to the Native American ways of life, the religious missionaries felt bound by their vocations to impose Christianity, as well as the "civilized" standards of the East, upon them. More often than not, the "missionaries" endeavors were rejected, leading to violent responses in a number of instances.

2.5   The US Army

As conflicts between the settlers and the Native Americans intensified, the US army was called in. Under President Polk, the Federal Government pursued an aggressive expansionist policy, encouraging western settlement in the belief that the more densely populated an area, the greater were America's claims to it. Hence, the intent was not only to protect the Pioneers but, as importantly, to stave off foreign claims and encroachment on what was perceived, by right of Manifest Destiny, to be legitimate US territory. To this end, military forts were set up along western rivers and the routes west. In addition to providing protection, these forts served as way stations, where pioneers could re-outfit their wagons and replenish themselves and their animals. When all else failed, territorial boundaries were settled by force, as in the case of the Mexican-American War of 1846–1848.

2.6   Cumulative Causation—The Year of Decision

Although the great migration west began in 1843, it did not achieve its peak until 1846—a year that De Voto describes as "The Year of Decision" (De Voto, 2019, 183). By then, positive feedback from those who had already gone west stimulated the fantasies of many Easterners, who, struggling on the margin to make a living, saw in the West a land of redemption (Dary, 2007, 80–83). One of the most influential interpreters of the West was Captain John Fremont, a topographical engineer. In a highly regarded book, he not only praised the virtues of the West, but also mapped it out, providing solid facts about the route, the signs, signals, and stopping places along the way (Dary, 2007, 83; De Voto, 2019, 80–81). Equally important in signifying the thriving Northwest were the writings and appeals of Nathanial Wyeth, a Cambridge businessman. Widely read and discussed by easterners hankering for a better life, his works became a standard primer for those contemplating a leap 2000 miles into an unknown territory (Billington and Ridge, 2001, 157–158).

There being strength in numbers, the effect of a growing number of westward expeditions was cumulative. As more and more settlers joined the caravans, the thresholds of reluctant travelers declined. The benefits of moving west now surpassed the risks entailed so that, by 1846, a tipping point was reached (Dary, 2007, 96).

2.7   The Dilemma of Collective Action

The gathering together of a large contingent of pioneers was only a first step in the migratory process. Even more challenging was the dilemma of collective action. Collective action problems—often referred to as 'the tragedy of the commons' or 'the prisoner's dilemma'—occur because people, when confronted with a choice of whether to pursue their own interest or the interest of a group as a whole, often favor themselves. Scholars have devoted considerable attention to this problem and how it might be resolved (Olson, 1971; Axelrod, 1984; Ostrom,

1990, 2000; Heckathorn, 1996; Carballo et al., 2014). Based on their analyses, we can stipulate that collective action will most likely be achieved when there are dense interactions leading to trust and reciprocal relations, the costs and benefits of participation are relatively equal, and the action is governed by local norms and rules established in a collaborative fashion from the bottom-up. *These conditions are most likely to come about when common standards can be agreed upon* (emphasis mine).

For the Pioneers, the problem of collective action was acute. Instead of a small group, made up of dense, trusting relationships, the pioneers comprised a large and diverse assortment of bold, adventurous, ambitious men who had little knowledge of one another. Although they shared a powerful common goal, their individual stakes could not have been greater, as they had with them their wives, children, and dependent relatives, as well as all their worldly possessions (Dary, 2007, 96). At the same time, the tasks of organizing themselves into an effective force able to withstand unforeseen menaces, creating an effective division of labor, and coordinating their behavior was immense (Gulati and Singh, 1998, 2). The Pioneers' solution was one that today's collective action scholars would certainly endorse—standard rules and procedures were set up in a deliberative process that led to a majority commitment (Dary, 2007).

The governance structure set up by the Pioneers not only stood them in good stead; it was also flexible enough to evolve as conditions changed. No sooner had they crossed the Mississippi River than they met together to establish traveling rules as well as a council of nine men to mediate disputes. Once they had crossed the Kansas River, they elected officers (Dary, 2007). The tasks and responsibilities of the travelers were clearly laid out. As one young Pioneer wrote in his diary:

> From six to seven o'clock is a busy time; breakfast is to be eaten, the tents struck, and the teams yoked and brought up in readiness to be attached to their respective wagons. All know when, at seven o'clock, the signal to march sounds, that those not ready to take their proper places in the line of march must fall into the dusty rear for the day.
>
> *(Dary, 2007, 84)*

On arriving in the Willamette Valley on May 2, 1843, the first settlers met to organize a government. Replicating the federal system, it was comprised of a legislature, an executive committee of three men, and a judicial system. On settling in their new land, however, they sought a more formalized, structured arrangement. To this end, they lobbied the Federal Government to incorporate Oregon into the United States. This desire for more formal governance reflects, in part, the changing balance between the costs and benefits of collaboration. For, as more and more settlers arrived, reciprocity and familiarity decreased. Moreover, with the common goal of Paradise having been achieved, and the dispersion of the settlers left to pursue their own individual goals, private benefits began to loom larger than collective ones.

2.8 Managing Complexity with Standards

With the deployment of the telegraph and the completion between 1851 and 1854 of the four major trunk lines linking the East and West, the Oregon Trail waned in prominence. Activities and interactions suddenly accelerated, greatly increasing the flow of people, goods, innovations, and investments to the West. Given these changes, the train was increasingly viewed as a symbol of national unity. At the same time, the increased flow of traffic and economic activity between the East and West greatly increased the complexity of the economy, giving rise to substantial transaction costs. To extract the full benefits of economic integration, these transaction costs had to be contained (North and Thomas, 1976). Standards provided a way.

2.8.1 Accounting for Transaction Costs

The concept of transaction costs refers to the time, energy, and money spent in gathering, processing, and employing economic related information (Williamson, 1985). The level of transaction costs is a function of uncertainty and the inclination and opportunity for economic actors to cheat one another. Such conditions are most likely to prevail when economic activities are carried out at great distances, when there are many different economic actors who, rarely interacting, are unknown to one another, when market information is impacted or unevenly distributed, and when production processes, worker skills, and products and services are highly differentiated and not substitutable for one another (Williamson, 1985). One way to reduce complexity, and thereby limit transaction costs, is to establish standardized products and procedures.

2.8.2 Reducing Transaction Costs with Standards

Following the great migration west, transaction costs were rife. On arriving in the West, the pioneers built homesteads, set up mining claims, established general stores, banks, and other small businesses. While located far from the teeming economic activity in the East, these western enterprises were dependent on their distant eastern counterparts for supplies, markets, customers, etc. In turn, eastern establishments sought to increase their gains by expanding their markets westward. But, although everyone benefited from trade, it was inhibited by a lack of market information and the uncertainties associated with doing business at a distance. Absent a communication infrastructure that could provide adequate east-west feedback, standards specifying product information and the means of exchange served to reduce uncertainty and thereby expand trade. As trade increased, so did the scope and intensity of interactions, and hence the need for additional standards.

2.8.3 Market-Based Standards

Not surprisingly, given this context, some of the most important standards developed early on were those related to the communication of market information and the mechanisms of exchange.[3] As the late James Beniger (1986) pointed out in his seminal work, *The Control Revolution: Technologic and Economic Origins of the Information Society,* during this period, market conditions and prices fluctuated widely from place to place. Lawlessness and opportunism were commonplace. To generate stable conditions and levels of trust essential for trade, standard economic processes and practices were required (OTA, 1992).

Consider the standardized roles of the middlemen who managed the trading process. They were central in this regard. Included among their roles was that of the commission agent, or factor, who carried out business on behalf of merchants in distant markets; the broker, who bought buyers and sellers together; the financiers, who provided a credit network to cover the up-front costs of transporting, processing, and distributing goods; as well as the retailers and other distributors, such as auctioneers and wholesale jobbers, the latter being of utmost importance in supplying western retailers (Beniger, 1986; OTA, 1992).

Equally important were the standardized trading forms and formats that helped to regularize trade by providing greater predictability (Beniger, 1986; OTA, 1992). Standard invoices, for instance, were used to document sales. The bill of lading was employed not only as a receipt but also as proof of ownership, as well as a negotiable instrument that could be traded for goods or used as collateral to back a loan. Catalogs also reduced transaction costs by displaying standardized products and listing their fixed prices (Beniger, 1986; OTA, 1992).

2.8.4 Institutionally Based Standards

Formal institutions, standardizing business practices, emerged as well. Included were common carriers such as the postal service and the railroads that—operating according to standardized procedures and a fixed schedule—allowed trade to take place on a consistent, periodic basis. Of equal consequence was the development of commercial law and legal precedents that standardized corporations, laid out a governing framework for interstate commerce, chartered insurance companies and commercial banks, and standardized ways of classifying products, to name but a few (Beniger, 1986; OTA, 1992).

By conveying east-west product information, 18th-century standards served not only to facilitate east-west trade; they also provided greater quality control. One of the first products to benefit from standards was food. Responding to scandals in the meat packing industry, Congress passed the Pure Food and Drug Act of 1906. This legislation not only protected against misbranding and food adulteration; it also standardized containers for marketing fruits and vegetables, thereby eliminating false measurements and deceptive shapes (Beniger, 1986; OTA, 1992).

When employed as trademarks, such quality standards increased the value of goods by allowing producers to differentiate their products from those of their competitors and to price products to different markets. American farmers played a major role in setting these standards, realizing that, by grading and classifying their products, they could establish separate distribution channels and increase their profits (Beniger, 1986; OTA, 1992). Thus, farmers labeled their products by their region of origin, while wholesalers used the names Goschen butter, Genessee flour, and Herkimer cheese as designations of their grade. By the end of the century, these quality standards efforts took the form of branding (Beniger, 1986; OTA 1992).

2.9 Failed Collaboration in the Railroads

The struggle to standardize the railroads illustrates how the lack of standards can inhibit collective action and the economic benefits to be derived therefrom.[4] Given high fixed costs, fluctuating demand, large-scale operations, and the need for coordination and specialized engineering skills, the railroad industry was always prone to high transaction costs (OTA, 1992). Nevertheless, they were unable to coordinate their operations, stabilize prices, and rationalize the industry structure. Railroad owners viewed cooperation as a zero-sum game: they assumed that any benefits accruing to a competitor was necessarily a loss for themselves, even when they had a joint interest in working together. At stake in their rivalry were the standards that controlled gateway routes, gateway processes, and gateway agencies.

Unable to come to terms, railroad magnates aimed to reduce their costs by merging and integrating their operations. Hence the mid-1880s witnessed the frenetic buying up of the industry's customers and competitors (Chernow, 1990; Chandler, 1993; Renehan, 2005). As Chandler (1993) described:

A multitude of commission agents, freight forwarders, and express companies, as well as stage and wagon companies, and canal, river, lake, and coastal shipping lines disappeared.

In their place stood a small number of large multi-unit railroad enterprises. By the 1880s the transformation begun in the 1840s was virtually completed.

To establish greater market stability, the railroads alternated their strategies between two extremes—cutthroat competition or pooling and price fixing. Because the economic uncertainties were so high, neither strategy proved successful. Cutthroat competition was ruinous for all, but cooperative agreements were untenable without some mechanism for enforcement (Chernow, 1990; Kennedy, 1991; OTA, 1992).

Moreover, although mergers reduced some of the uncertainties facing companies from their external environment, they greatly exacerbated problems of internal control (OTA, 1992). Safety problems were a major concern. To rationalize their internal operations, the railroads adopted standardized role assignments and procedures as well as innovations such as through bills of lading, standardized cars, uniforms, track gauges, standardized couplers, and air brakes. To deal with these new administrative complexities, railroad companies developed an entirely new type of business enterprise, based on a standardized bureaucratic and hierarchical organizational structure. The railroad model, which others quickly emulated, prepared the way for an entirely new industrial era (Beniger, 1986; OTA, 1992).

Notwithstanding these radical changes, all transaction costs did not disappear. On the contrary, as is often the case, efforts to reduce transaction costs in one part of a system lead to the emergence of new economic bottlenecks in other parts, and then to new government efforts to control them. To reduce the uncertainty stemming from over building, railroad magnates pushed their social, political, and economic advantages to the limit.

Railroad owners were not alone in questioning whether the market could solve the problem of collective action functioning on its own. The railroads were at the center of national economic activity. The nation's financial markets were greatly influenced by rail financing, and commodity prices were directly linked to railroad rates (Knowles, 1967) and railroad. Not surprisingly, the bankers were the first to intervene. To force a solution, J. P. Morgan—acting as a neutral third party—sequestered many of the key railroad owners on his yacht. Acting under duress, the railroad moguls were able to come to terms. However, they quickly reneged on their agreement, there being no means of enforcing it (Chernow, 1990; OTA, 1992).

The railroad owners' competitive machinations quickly spilled over into the political arena. Most vocal in calling for reform were small business owners and farmers in the West who had been forced by the railroad companies to subsidize the discounted rates offered to the large, eastern industrialists. An increasingly disgruntled and activist labor force soon joined these forces. Under mounting pressure, the government decided it was time to intervene. Acting to protect consumers and stabilize the market, the government established the Interstate Commerce Commission in 1887. Three years later, Congress passed the Sherman Antitrust Act—a law intended to inhibit monopoly power (OTA, 1992).

2.10 Expanding the Pie

The interaction and exchange between East and West increased the fitness of the nation's economy as a whole. As economic historian Douglass North has pointed out, the period between 1840 and 1880 was not only a time of great land acquisition; it was also an occasion of major economic restructuring (North, 1966, 205–215). Due, in part, to the Western demand for manufactured goods, the Northeast became a market based manufacturing center. Moreover, economic efficiency in the production of manufactured goods grew as a result of a deeper division of labor and increased economies of agglomeration. As the region developed, it became a growing market for western agricultural goods. At the same time, agricultural productivity in the West greatly increased due to a growing population, increased eastern investment, higher agricultural prices due to growing demand, and the deployment of agricultural technology imported from the East (North, 1966, 205–215).

Of course, one cannot judge the costs and benefits of the Oregon Trail by these economic factors alone. The economy does not exist in a vacuum; it is integrated into the "whole" by networks of all kinds—social, political, and cultural. Thus, we see that as the population in the Oregon country increased, its inhabitants imported governmental models from the East. In turn, the US government incorporated Oregon as a national territory in 1848 and as a state in 1859.

As importantly, the interaction and exchange between the East and the West enhanced the nation's culture, meshing together diverse memes, artifacts, symbols, signs, and yarns into a unified blend. The advent of inexpensive popular magazines, such as *The Saturday Evening Post, The Ladies Home Journal,* and *Country Gentleman,* as well as product catalogs such as *Sears* and *Montgomery Ward*, served to standardize tastes from East to West. At the same time, standard stories of western heroes and western ways of life became the fodder for dime store novels. Consider, for instance, the wild western men and women such as Seth Jones, Deadwood Dick, Calamity Jane, and Denver Doll, as they were portrayed in the East (Smith, 1990, 91–120).

2.11 The High Costs Entailed

Notwithstanding these benefits, the costs of building the Oregon Trail were also high. One need only ponder, for example, the death of the 30,000 pioneers who never achieved "the promised land," the death and mistreatment of so many Native Americans along with the destruction of their homelands and culture, the devastation of the land due to the railroad boom, and the speculation and bank failures associated with railroad building. And, of course, there was the cost of securing the Oregon Territory, especially the engagement in the Mexican-American War, which not only entailed considerable financial outlays and loss of lives but also led to the long-term deterioration of the US relationship with Mexico.

The story of the Oregon Trail illustrates the layered process by which actors in complex environments can make their way up to a higher level of fitness, given cooperation and the sharing and integration of standard protocols. It shows, moreover, how—at higher levels of the fitness landscape—evolving standards give rise to greater specialization and differentiation, as well as the divergence of actors' interests. Hence, whereas at lower levels of fitness, actors can more readily interact to find common ground, at high levels of fitness, standards will likely generate winners and losers. It is, thus, at higher levels of fitness that bottom-up emergent standards process typically give way to third party, top-down approaches.

Notes

1 The terms "fitness levels" and "fitness landscapes" were developed by Sewall Wright.
2 As Seven Johnson opines, "The adjacent possible is a kind of shadow future, hovering on the edge of the present state of things, a map of all the ways in which the present can reinvent itself. Yet it is not an infinite space or a totally open playing field. The number of first-order is vast, but it is a finite number, and it excludes most of the forms that now populate the biosphere. What the adjacent possible tells us is that at any moment the world is capable of extraordinary change, but only certain changes can

happen ... The strange and beautiful truth about the adjacent possible is that its boundaries grow as you explore those boundaries" (Johnson, 2002, 17–18; see more generally, 14–25).

3 The following sections, from pages 20–26, are drawn from my 1992 OTA report, *Global Standards: Building Blocks for the Future*, in which the work of James Beniger(1986) is extensively cited.

4 This section on the railroads draws heavily on the 1992 OTA report, *Global Standards: Building Blocks for the Future.*

References

Applegate, Shannon, and Terence O'Donnell (1994) *Talking on Paper: An Anthology of Oregon Letters and Diaries*, Corvallis, OR: Oregon State University Press.

Axelrod, Robert (1984) *The Evolution of Cooperation*, New York, NY: Basic Books.

Beck, Warren A., and Ynez D. Haase (1989) *Historical Atlas of the American West*, Norman, OK: University of Oklahoma Press.

Beckham, Stephen Dow, eds. (2006) *Many Faces: An Anthology of Oregon Autobiography*, Corvallis, OR: Oregon State University Press.

Beinhocker, Eric D. (2007) *The Origin of Wealth: The Radical Remaking of Economics and What It Means for Business and Society*, Cambridge, M: Harvard Business School Press.

Beniger, James (1986) *The Control Revolution: Technological and Economic Origins of the Information Society*, Cambridge, MA: Harvard University Press.

Billington, Rae Allen, and Mark Ridge (2001) *Western Expansion: A History of the American Frontier*, 6th edition, Albuquerque, NM: University of New Mexico Press.

Boone, R T. and Macy, M.w (1999) "Unlocking the Doors of the Prisoner's Dilemma: Dependency, Selectivity and Cooperation," *Social Psychology Quarterly* 62 (1), 32–52.

Brown, John Seely, and Paul Duguid (2017) *The Social Life of Information*, Cambridge, MA: Harvard Business School Press.

Carballo, David M., Paul Roscoe, and Gary M. Feinman (2014) "Cooperation and Collective Action in the Evolution of Complex Societies," *Journal of Archeological Method Theory* 21, 98–133.

Chandler, Alfred D. (1993) *The Visible Hand: The Managerial Revolution*, Cambridge, MA: Harvard University Press.

Chernow, Ron (1990) *The House of Morgan: An American Banking Dynasty and the Rise of Modern Finance*, New York, NY: Grove Press.

Dary, David (2007) *The Oregon Trail: An American Saga*, New York, NY: Alfred A. Knopf.

De Voto, Bernard (2019) *The Year of Decision*, Pickle Partners Publishing, www.pp-publishing.com

Eggertsson, Thrainn (1991) *Economic Behavior and Institutions*, New York, NY: Cambridge University Press.

Ernst, Dieter, and Linsu Kim (2001) *Global Production Networks, Knowledge Diffusion, and Local Capability Formation: A Conceptual Framework*, Paper presented at the Nelson & Winter Conference in Aalborg, Denmark, Honolulu, East-West Center.

Granovetter, Mark (1987) "Threshold Models of Collective Behavior," *The American Journal of Sociology* 83 (6), 1420–1443.

Gulati, R., and H. Singh (1998) "The Architecture of Cooperation: Managing Coordination Costs and Appropriation Concerns in Strategic Alliances," *Administrative Science Quarterly* 43 (4), 781–814.

Hardin, Garrett and John Baden (1998) *Managing the Commons*, Bloomington, IN: University of Indiana Press.

Heckathorn, Douglas D. (1996) "The Dynamics and Dilemmas of Collective Action," *The American Sociological Review* 61 (2), 250–277.

Hine, Robert, John Mack Faraghar, and John T. Coleman (2000) *The American West*, New Haven, CT: Yale University Press.

Holland, John (1999) *Emergence: From Chaos to Order*, New York, NY: Basic Books.

Holland, John (2014) *Signals and Boundaries: Building Blocks for Complex Adaptive Systems*, Cambridge, MA: MIT Press.

Johnson, Steven (2002) *Emergence: The Connected Life of Ants, Brains, Cities and Software*, New Yor, NY: Simon and Schuster.

Johnson, Steven (2010) *Where Good Ideas Come From: The Natural History of Innovation*, New York, NY: Bloomberg.

Kauffman, Stuart (2008) *Reinventing the Sacred: A New View of Science, Reason and Religion*, New York, NY: Perseus Group.

Kennedy, Robert Dawson Jr. (1991). The Statist Evolution of Rail Governance in the United States 1830–1986, in Campbell, L., J. Rogers Hollingsworth, and Leon Lindberg, eds. *Governance of the American Economy*, New York, NY: Cambridge University Press.

Kesselman, Amy (1974) *Diaries and reminiscences of women on the Oregon Trail: A study in consciousness, Portland State University, PDXScholar Dissertations and Theses*.

Knowles, L.C.A. (1967) *Economic Development in Nineteen Century France, Germany, RUssia, and the United States*, New York, NY: Augustus M. Keller Publishers, Reprints of Economic Classics, 91–93.

North, Douglass (1966) *The Economic Growth of the United States*, New York, NY: W.W. Norton & Company.

North, Douglass and Robert Paul Thomas (1976) *The Rise of the Western World: A New Economic History*, New York, NY: Cambridge University Press.

Olson, Mancur (1971) *The Logic of Collective Action: Public Goods and the Theory of Groups*, Revised edition, Cambridge, MA: Harvard University Press.

Ontko, Andrew Gale (1997) *Thunder Over the Ochoco, The Gathering Storm, Volume 1*, Bend, OR: Maverick Publication.

Oregon Interpretive Center, La Grande, Oregon, visit, 2018.

Ostrom, Elinor (1990) *Governing Commons: The Evolution of Institutions for Collective Action*, Cambridge: Cambridge University Press.

Ostrom, Elinor (2000) "Collective Action and the Evolution of Social Norms," *Journal of Economic Perspectives* 14 (3), 137–158.

OTA, US Congress (1992) *Global Standards: Building Blocks for the Future*, Washington DC: US Government Printing Office.

Pitzer, Erik, and Michael Affenzeller (2011) "A Comprehensive Survey of Fitness Landscape Analysis," in Fodor, J., Klempous, R., Suárez Araujo, C.P., eds. *Recent Advances in Intelligent Engineering Systems*, Berlin: Springer.

Polanyi, Karl (1962) *Personal Knowledge: Towards a Post Critical Philosophy*, Chicago, IL: University of Chicago Press.

Quillen, Carol Everhart (2023), "Humanism and the Lure of Antiquity," in Najemy, John M., eds. *Italy in the Age of the Renaissance*, New York, NY: Oxford University Press.

Renehan, Edward J. Jr. (2005) *Dark Genius of Wall Street, The Misunderstood Life of J. Gould*, New York, NY: Basic Books.

Roosevelt, Theodore (1889) *The Winning of the West*, New York, NY: G. P. Putnam and Sons.

Smith, Henry Nash (1990) *Virgin Land: The American West as Symbol and Myth, 21st Printing*, Cambridge, MA: Harvard University.

Vandenbroucke, Gillaume (2008) "The Western Expansion," *International Economic Review* 49(1), 181–121.

Williamson, Oliver (1985) *The Economic Institutions of Capitalism*, New York, NY: The Free Press.

PART II

History through the Lens of Complexity Theory

3

STANDARDS, NORMS, AND EMERGENCE IN SOCIAL SETTINGS

3.1 Analyzing Emergence

Emergence requires rules to structure interactions[1] (Holland, 1995; Mayfield, 2013, 7). These rules (which I characterize as standards) allow system elements—such as agents, actors, and components—to generate complex outcomes through their interactions, giving rise to expanded opportunities and new rules leading to greater complexity and higher fitness levels. At the same time, these rules serve as social facts[2] with respect to entities lower down the fitness landscape. They serve to inhibit some interactions, allow for others, as well as determine how interactions take place (Sawyer, 2005). However, as detailed in the following chapter, when fitness landscapes shift appreciably, existing standards lose their values, so systems collapse in a phase transition,[3] providing opportunities for new, lower-level actors to generate additional resources and ferment emergence with restructured standards configurations.

Quantitative analysts and computer scientists have modeled these processes in their efforts to document and extend the theory of complexity. Their models have incorporated rules by embedding them in computer programs in the form of if/then statements. By running their models over time, complexity theorists have been able to trace how rules generate emergence and the outcomes associated with it. Because computer scientists can posit diverse rules in a top-down fashion, they are able to make real-time comparisons about how outcomes change with respect to changing rules (Marro, 2014, 4–9). What they cannot do very well, however, is to account for how rules emerge, or how historical contingencies might influence their evolution.

Social scientists face a far greater challenge in accounting for emergence. To begin with, one barrier to their efforts is their disagreement about the ontology of the social world and the appropriate methodology for analyzing it (Bunge, 2003, 130; Demeulenaere, 2011). Even before sociologists address the subject of emergence, they must first theorize the very existence of rules, be they in the form of social facts, norms, habitus, manners, etc. (Bunge, 2003, 190; Elder-Vass, 2010). As significantly, in contrast to quantitative analysts, most social scientists must ascertain, qualitatively, how standards evolve over time, in diverse situations and settings. For instance, they must deal with questions such as: Do social standards emerge through local interactions in a phenomenological fashion, or are they prescribed through more deeply structured,

DOI: 10.4324/9781032721125-6

institutional processes? Social scientists must also determine whether standards, taking the form of norms, constitute dominant social structures or whether—and the extent to which—actors have agency to reformulate and reconstruct social standards in the course of their enactment.[4] How these standards are enforced, and by whom, must also be addressed.

Unfortunately, wide-ranging, and often irreconcilable, perspectives with respect to these issues are prevalent throughout the social sciences (see, for instance, Alexander, 1983; Bunge, 2003; Elder-Vass, 2010; Layder, 2015). Viewpoints range from the theoretical, systems-based traditions of Parsons and Marx to the phenomenalist approaches that focus on social interaction from the individual actor's point of view, as in the approaches employed by Garfinkle (1967) and Berger and Luckmann (1967). Midway between these two approaches are those developed by Giddens and Bourdieu, which seek to conceptualize societal interactions based on the interplay between individual agency and structural constraints (Stones, 2005; Atkinson, 2020).

One way of transcending these debates is to consider standards from a complexity perspective. Complexity theory views systems as being layered hierarchically, such that agents must—in order to move up the fitness landscape, or outward to an adjacent possible—adapt to, and develop new standards appropriate to each level. Moreover, standards emerge and take on distinct forms depending on the context in which they manifest themselves. When viewed collectively, standards generate networks that serve as platforms upon which new activities, as well as novel standards to accommodate them, arise. In fact, it is upon these platforms that standards provide a path for actors or agents to leap from one layer of the complexity hierarchy to the next, opening the way for social evolution and the emergence of a "whole."

Accordingly, to identify the common features of emergence, and how it gives rise to greater fitness, I compare the process of emergence in a number of very diverse settings. To accommodate these differences, I view each case as a separate field—much as in Bourdieu's conception of fields—so that each operates according to its own standardized rules of interaction as well as its own ways of generating these rules (Hilgers and Mangez, 2014; Atkinson, 2020, 81–120). However, in contrast to Bourdieu, I view fields not primarily as spaces of contention for dominance but also—and as importantly—as platforms for integration and collective action. For what occurs on these platforms is essentially a function of existing norms. In this respect, I view norms, as characterized by Durkheim (2014) and Parsons (2013), as the rules[5] of interaction that lead to emergence both within and beyond the field. As well, I consider what happens when norms fail to emerge, or are ruptured by examining the case of today's Republican Party, which has become so mired in its own wrangles and disagreements that it has been unable to generate any emergent outcomes at all.

3.2 Stuart Kauffman's Search for Order in Complexity

Years ago, during a long flight from DC to Boise, I found an intriguing book to keep me engaged. This book, *At Home in the Universe: The Search for the Laws of Self Organization and Complexity* (1995), written by complexity biologist Stuart Kauffman, was hardly one of those best-selling mysteries typical airport fare. Nonetheless, it was a thriller. Challenging the long-held paradigm proffered by Darwin, which posits that evolution is an incremental process driven by a battle among the fittest for survival, Kauffmann contended that the evolutionary process was, at times, a very bumpy path entailing far more than natural selection. According to Kauffman, while Darwin's account of evolution explained how natural selection might cull the weakest candidates from a pool of contestants, allowing winners to propagate over time, it left unanswered the question of how the pool of contenders, itself, might have—like in the Cambrian period[6]—greatly expanded in size and complexity in a relatively short period of time. Kauffman's answer was groundbreaking. He claimed that, although natural selection winnowed the pool of nature's survivors, it did so not only by selecting from the progeny of former survivors but also by choosing from among a vastly larger and more diverse reservoir of life forms that spontaneously emerged, endogenously, from within the life process itself. To Kauffman, the viability of self-organized life forms was evidence of some kind of order, bordering on sacredness, in the universe.

Kauffman has employed this argument in his critique of scientific reductionism, a theme he has pursued over the years. Underlying his acerbic condemnation of reductionism is his disdain of physicists' assertions that knowledge—even an understanding of life—can only be obtained by breaking down phenomena into their fundamental parts and deducing their wholeness and essence from the behavior and interrelationships among them. In his writings, Kauffman rails against this claim, asserting that biology cannot be reduced to physics.[7] Life, he contends, is an emergent phenomenon that, drawing upon the creativity of the universe, can independently generate and sustain itself. According to Kauffman, by denying life any agency of its own, reductionism precludes values, intent, creativity, and the spirituality that gives meaning to life.[8]

Reading Kauffman's book, I was compelled to ask myself where I might find other instances of emergence. The answer, I found out, is just about everywhere.

3.3   The Emergence of Complexity Science at the Santa Fe Institute

Kauffman was not alone in his dissatisfaction with the reductionist direction of science. There were, at the time, a number of scholars who found themselves stymied in their research and careers by traditional gatekeepers insisting on a reductionist approach. For one, Brian Arthur, a tenured professor at Stanford University, was typically rebuffed by his colleagues in the economics department when he tried to engage them in a discussion of increasing returns[9]—a phenomenon he uncovered that undermines the concept of equilibrium, a fundamental postulate in neoclassical economics (Waldrop, 1992, 13–60). Similarly, John Holland, a formidable computer scientist and mathematician at the University of Michigan, found little support and enthusiasm among his peers for his efforts to develop a genetic algorithm that could track the process of emergence by observing agents that learned, adapted, and mounted the hierarchical building blocks leading to a higher fitness landscape.[10] What, if anything, his colleagues asked, do genetic algorithms have to do with computer science?

The tale of how these innovative, but often rebuffed, scholars found common ground in complexity science is the story of the Santa Fe Institute. For they were not the only ones troubled by reductionism. The former head of research at Los Alamos, George A. Cohen, also pondered whether existing approaches to scientific inquiry were adequate to address the increasingly complex issues of the future. He called for a more holistic approach, one that transcended traditional disciplines with their restrictive methodologies. From Cohen's perspective, to develop such an approach required stepping out of the academic milieu where incentive structures reinforced academic silos. Instead, Cohen called for rising above the traditional research establishment and creating a new institute whose mission would be nothing less than developing a new science for the 21st century (Waldrop,1992).

Surely, this was a transformative vision. But a big question remained: how to bring it about? To gain credibility and support for such an institute, Cohen reached out—with the help of the highly respected and well-connected physicist, Murry Gellman—to scholars renowned in their fields, a number of whom were Nobel Prize winners. A limited amount of funding for the Institute was initially made available through the National Science Foundation and the Department of Energy, but the sizable funding necessary to provide ongoing support only came when John Reed at Citibank agreed to fund a three-year effort to explore the implications of complexity for the economy. Bringing all the necessary pieces together was a slow, trudging process, but by 1984, the Institute was finally incorporated and ready to go (Waldrop, 1992; Lewin, 1999).

Significantly, the *modus operandi* that emerged at the Santa Fe Institute was a microcosm of complexity itself (Waldrop, 1992; Lewin, 1999). It was a bottom-up, interactive process. Apart from a broad mandate to focus on complexity and the economy, there was no top-down directive mapping out what issues to cover and how to proceed. Thus, when Brian Arthur was tasked to carry out the economic project, Cohen left decisions as to how to move forward essentially up to him. As it turned out, given this open-ended environment, and as complexity theory might predict, the Institute's highly successful process of investigation emerged on its own through discussions and interactions among the participants intermingling at Santa Fe.

These interactions were key to the Institute's success. Participants were of a high caliber, having been selected based on their reputations, accomplishments, diverse backgrounds, intellectual interests, and enthusiasm for sharing their ideas and gaining feedback from others. Typically, when engaged in workshops oriented around specific topics or intellectual queries, they cross-fertilized each other's ideas, from which new, unforeseen topics from the adjacent possible emerged.[11] Informal interactions were also the norm, as scholars routinely met for coffee, lunch, or a walk in the mountains to test and exchange their ideas.

These relationships were not limited to the Institute. Like the "autocatalytic sets"[12] that Stuart Kauffman had defined, these connections took on a permanence, which extended to scientific endeavors elsewhere. What made such intellectual networks possible was the existence of widely available, advanced computers and the many types of computer simulations that served as common languages to communicate across projects and disciplines[13] (Waldrop, 1992). It was indeed the study of complexity at the Santa Fe Institute that fostered a growing interest in the phenomenon of emergence and the complex outcomes to which it gives rise.

3.4 Cross-border Norms at the Arava Institute

Many years ago, I accompanied my husband Brock on a great adventure at Kibbutz Ketura in the Israeli Negev Desert. It was here, on the kibbutz, that my husband had agreed to teach courses on environmental law and politics at the Arava Institute for Environmental Studies. Established in 1996 following the Oslo Accords, the Arava Institute aims to promote peace and environmental sustainability in the Middle East through trans-border collaboration on educational programs, research undertakings, and environmental projects.[14] The students in the program were very diverse: they included Jews and Christians from the United States, Israeli Jews and Arab Christians, Palestinian and Jordanian Arabs, as well as Hyatt, a student from the Druze community in the Golan Heights. Our students had entered the program not only with their own, individual aspirations and goals but also with some personal biases and resentments. Compounding the challenge of living together with the "other," the students had to negotiate their differences in a very unique cultural environment, that of life on kibbutz.

While focusing on common concerns about the environment, the Institute aimed simultaneously to promote peace through collaborative interactions. Thus, to connect students around environmental issues that were relevant to all, the Institute had to induce norms changes, shifting the atmosphere away from diffidence and hostility to norms centered on greater tolerance. This, of course, was no minor task! To set the stage and establish a platform for productive interactions, students were provided with joint living arrangements, sympathetic mentors, common experiences, a framework for dialogue, and a wide range of information and activities related to their shared environmental concerns. Notwithstanding smoldering background tensions, things were off to a pretty good start at the beginning of the semester. However, midway into the term, and just as we were readying ourselves for a bus trip to the north, antagonisms began to mount. The first stop on our trip was Majdal Shams,[15] a Druze village in the Golan Heights, from where—beyond a barbed wire fence, and across a very large gully—we could see Syria. Here was the birthplace of Hyatt, and it was here, at her parent's home, that we were invited to share a meal. However, it was here, too, that tempers boiled over. For reasons I can no longer remember, our Israeli Jewish students refused to come to dinner, remaining instead on the bus.

This was a major turning point. The ethnic animosities among our students could not go unchecked and unresolved. It was time to bring tensions to the foreground and negotiate new ways of interacting, such that our students might agree to them. Thus, throughout that evening, we met in our hostel where everyone shared their pent-up feelings and concerns. As it turned out, the gathering was one of the most productive standards-setting meetings I have ever witnessed. Having a common interest in making the most of their time at the Institute, students worked out their differences, agreeing in the end of a night of heated exchange on a *modus operandi* to which they could all commit. What emerged from this deliberation was a new level of collegiality and an eruption of enthusiasm on behalf of the Institute's goals. The following morning, we all piled on our bus, bustled down Highway Six, and arrived in Tel Aviv, where, at a meeting with Israeli Prime Minister Simon Peres, we presented a united appeal on behalf of the Middle East environment.[16]

When standards of cooperation are reenacted time and again, they become, like a well-worn path, part of the background. This explains why, even at a time when ethnic conflict is raging in Israel, the Arava Institute remains a vibrant oasis in a desert of hostilities.[17]

3.5   Jamming, and All That Jazz

What does "a new science for the 21st century" have to do with jazz? Apparently, a great deal! Just as the emergent outcomes at the Santa Fe Institute are based on improvisation and adaptation, so too are the performances of jazz bands and improv theater productions. Both genres conform to the path of emergence anticipated by complexity theorists. For instance, jazz ensembles and improv productions both start out from scratch, with no prior scheme or direction. The theme, or plot, of a presentation emerges collectively through the interaction of multiple participants, so that what ultimately emerges is totally unpredictable both to the participants and the audience.

The process of emergence in improvisation is the product of interaction governed by a limited number of conventions and rules. Understanding and committing to follow these rules is a prerequisite for participation and engagement. Thus, joining a jazz band or improv performance and adhering to these standards is entirely voluntary: as we saw in the case of the

Arava Institute, it results from a desire to be part of the whole. Most notably, in the course of interactions, new artistic forms emerge as well as new rules of interaction. As described by Sawyer (2008):

> In jazz, for example, no single performer can determine the flow of the performance: It emerges out of the musical conversation of a give-and-take as performers propose new ideas, respond to other's ideas, and elaborate and modify these ideas as the performance moves forward. (4)

Likewise, in improv theater productions, emergent outcomes are highly complex. As Sawyer (2008) notes:

> Actors cannot know how their turns will be interpreted by others; each turn gains its final meaning only from the ensuing flow of discourse.... each turn of dialogue, although spoken by a single actor, eventually takes on a dramatic meaning that is determined by a collaborative, emergent process. (6–7)

As notable, what emerges from improvisations is not only a performance for the audience but also, and as importantly, an altered state for the performers—a condition that Mihaly Csikszentmihalyi characterizes as *flow* (Csikszentmihalyi, 1990). Much as we see with a flock of birds, all flying in a uniform trajectory, so too a group of performers all act as one when in a state of flow; "each performer is open and listening to the others, and each performer fully attends to what the others are doing, even as they are contributing to the performance themselves" (Sawyer, 2008, 45). The intensity of these interactions becomes so great that performers extend themselves beyond their limits, and in so doing achieve a state of euphoria and bliss (Csikszentmihalyi, 1990).

It must be emphasized, however, that notwithstanding the open-endedness and unscripted nature of such endeavors, improvisation requires some constraints. Much as we learn from complexity science, lacking some fundamental standards of interaction, performances would be far too chaotic. On the other hand, if standards are too stringent, performances will become frozen in time and place. Ideally, then, the rules governing improvisations should be located in what Kaufmann describes as the "realm of complexity," where, as he points out, is to be found at the 'edge of chaos' (Kauffman, 1995, 43).

Not surprisingly, then, it is here, at the edge of chaos, where the rules of jazz and improv theater reside. Participants in these domains must be able to innovate, but to do so they must build upon shared conventions and repertoires, which—like a shared language—allow group members to lead and follow one another as a unified whole. As Sawyer (2005) contends:

> Although actors must learn these conventions before they can engage in effective improvisations, the conventions are not so much constraining as they are enabling—they enhance the creativity of the group and increase its improvisationality. Thus the conventions do not have the effect of making improvisational performances more structured and scripted: they paradoxically have the opposite effect, of making the performance more collaborative, more improvised, and more emergent. (54)

The cases of jazz bands and improv productions illustrate an essential point with respect to sociological debates about structure and agency. Structure, when taking the form of norms, is not necessarily an adversary of agency. In fact, as described above, norms, even in the form of constraints, can enable and reinforce actions that lead to emergence. As importantly, as detailed below, the absence of shared norms typically leads to chaos and the inability of action to take place.

3.6 Emergence in the Italian Renaissance

Like the members of a jazz group, the inhabitants of Italy in the 13th century had to improvise. For it was the first time since the fall of Rome that Italians had been on their own, void of foreign invaders. At one stroke, the Italian peninsula was split into a number of diverse territories, each with its own history and traditions. It was, I contend, these loosely structured interactions among Italy's newly formed competing polities that led to the emergence of the Renaissance and the flow of creativity accompanying it. Many historians do not agree. Focusing solely on isolated facts and events, they fail to account for the Renaissance as a whole, that is, as it being greater than the sum of its parts. To capture the essence of the Renaissance, a complexity perspective is key.

3.6.1 Experiencing Florence

One way to capture the ethos of the Renaissance is to experience it in person, with a trip to Italy, where one can find the lingering residue of *flow*. Much like the humanists, who sought out distant places in search of ancient wisdom and eloquent Latin, I travelled to Florence as a student to enhance my education not only by studying art history and Italian politics but also by perfecting my Italian. It was a transformative experience, as I emerged a far more enlightened person.

Studying in the heart of the Renaissance took place in real-time. It was a conjunction of the past and the present. History and art surrounded me: I was immersed in it. On any day, I might leave my art history class and stroll the ins and outs of a narrow, cobbled street, only to stumble on an exquisite Andrea del Sarto fresco beckoning me from a remote corner wall. Other days, I might learn of some important historical event, and then—a few moments later, and a short distance away—explore the exact site where it had occurred. As telling, because my school abutted the Piazza Savonarola, I was reminded daily of the charismatic, authoritarian priest who, having become the tyrannical ruler of Florence, was burned to death in the Piazza Della Signoria.

3.6.2 Did the Renaissance Really Happen?

Living in the midst of such historical remembrances and visual stimuli, I experienced—if only vicariously—the flow of the Italian Renaissance. You can imagine my surprise, then, to find that many recent historical accounts of the late Middle Ages (13th–16th centuries) make little mention, much less any explanation, of the Italian Renaissance (Celenza, 2006; Ruggiero, 2015, 29–30). Some have alleged that the Renaissance was but a continuation of the late Middle Ages and thus need not be the subject of special attention. Others, pointing to the multiple scientific advancements in the 12th century, contend that the Italian Renaissance was far from unique (see, Benson et al., 1982). Moreover, some historians have even derogated the Renaissance by claiming that its aura was contrived by its self-interested participants. As Brown (2021) points out:

> [Presently] … few would interpret [the Renaissance] uncritically as optimistic new birth. Instead, they would call it a piece of publicity or propaganda written to praise not only Florence… but also the Medici family, who were its own patrons and helped to promote this cultural revival. (9)

This kind of dissension is understandable. Generally speaking, a new cadre of scholars will put forward revisionist ideas that challenge existing paradigms (Celenza, 2006; Kuhn and Hacking, 2012). And to be sure, many of those downplaying the Renaissance have reacted to a long-established and well-entrenched paradigm created by the 19th century Swiss historian Jacob Burckhardt.

In his opus, *The Civilization of the Renaissance in Italy*, published in 1860, Burckhardt characterized the Renaissance in terms of what we might describe today as a *phase transition*—the subject of the following chapter. That is to say, Burckhardt viewed the Renaissance as an emergent cultural phenomenon, which paved the way for modernity and independent, empowered individuals who had considerably more agency over their lives than those living in the Middle Ages. Because Burckhardt employed a broad cultural lens, which cut across events in a number of domains, he was able to link Renaissance episodes in complex ways by emphasizing the role of humanism as a common language and set of standards. But ironically, it was Burckhardt's accentuation of humanism that provided an opening for many of the intellectual challenges that followed.

With these criticisms in mind, intellectual historian Christopher Celenza has, like me, pondered what happened to the Italian Renaissance. Where did it go, he asks? In his response, *The Lost Italian Renaissance; Humanists, Historians and Latin's Legacy*, Celenza argues that it is the profound misreading of humanism[18] that accounts for the failure of historians to recognize the unique flowering and impact of the Renaissance (Celenza, 2006; Maxson, 2013). As he describes, early on in the post-Renaissance period, philosophers—ensconced in the prevailing German metaphysical school—asserted that humanism was irrelevant to Italian history because Latin and the classical writings from the past were divorced from Italy's population as a whole. In fact, as he says, these critics went so far as to claim that the humanists' focus on classical Latin would undermine Italians' ability to develop a literature of their own (Celenza, 2006). This belief that language was a derivative of national culture, which prevailed throughout the 19th century, perverted the way Renaissance humanism was conceived. As Celenza (2006) observes:

> … Renaissance Latin was never even seen as a possible field of serious inquiry for many European intellectuals. First, there existed a belief inherited from the Enlightenment regarding language, to wit, that only a native tongue could truly express the essential genius of a people. When this assumption regarding language merged with the second factor, the rise of nineteenth-century nationalist historiographies, Renaissance Latin was doomed. (2)

American scholars, writing about humanism in the post-WWII era, were susceptible to these criticisms. To skirt the controversies, they defined humanism so as to neutralize it, denying it agency by disembodying it from any intellectual or philosophical content. Most prominent in this regard was Oskar Kristeller, who redefined humanism as a new, more comprehensive educational paradigm based on the humanities. As Celenza explains, "Kristeller could never countenance the notion that, by self-consciously turning away from metaphysics, some humanists were actually making a philosophical statement" (Celenza, 2006, 51).

In evaluating the work of Burckhardt and his interpretation of the Renaissance, it is important to keep in mind that, of all the historians contemplating the Renaissance, he alone generated a paradigmatic construct of it.[19] As Ruggiero has argued, this is precisely what is missing from our understanding of the Renaissance (Ruggiero, 2015, 29). To provide this missing dimension, which encapsulates and accounts for the totality of the Renaissance, the place to start is not, as recent historians have done, in reductionist explanations that focus on the participants and their productions, but rather, much more as Burckhardt might have done, building on complexity theory and the structural variables that fostered the multitude of interactions that engendered the Renaissance emergence.

3.6.3 The Emergent Renaissance

Complexity theory provides a useful way to encapsulate the Renaissance in a holistic framework that not only incorporates the players and events that constituted the period under consideration but also, and more importantly, illustrates how these happenings were joined together to generate emergence. For emergence to take place on a large-scale, a number of conditions are typically required. First, there must be a sizable group of heterogeneous actors involved. Second, these actors must have the requisite status/resources to participate in a common undertaking. Third, actors must have a common space—as in a place, network, or platform—in which they can channel their interactions. Finally, actors must adhere to a common set of standards—as in a language or code of behavior—according to which they can communicate and impact the behavior of others. In the 13th century, Italy was ideally situated to meet these requirements.

Heterogeneity, for one, was a prominent characteristic of Italy throughout the Renaissance. Although Italy had suffered like the rest of Europe from the plague and lost almost half of its population, it was not long thereafter—in the latter half of the 13th century—that Italy's population mushroomed, increasing the diversity of its inhabitants as well as the division of labor (Najemy, 2005; Brown, 2021, xxiv). As importantly, in contrast to other European entities, where individuals were agglomerated into burgeoning, autocratic states, Italy's swelling population was cobbled together from a wide variety of autonomous, organized groups and networks, which competed with one another for prominence[20] (Najemy, 2005).

As a result, diversity prevailed at every level of Italian society. At the local level, for example, artisans and professionals organized themselves into a broad array of specialized and highly competitive self-governing guilds (Ruggiero, 2015, 90–94).[21] At the regional level, members of prominent families operated as competing private corporate bodies, led by strong men, who ruthlessly sought to eliminate their opponents and dominate the political process.[22] Likewise, across the Italian peninsula, church leaders from Guelf and Ghibelline factions battled each other, while the nobility and the *popolo* fiercely contended with each other for status and power.[23]

Then, even as consolidation followed, conflict among diverse partisans prevailed at the higher, regional level. In a chain reaction, elites—aggrandizing their claims and transcending their individual constituencies—engaged in major conflagrations resulting in the amalgamation of competing regional territories dominated by powerful warlords who derived their standing from their ancient lineages and military prowess[24] (Martines, 1979; Najemy, 2006; Bartlett, 2019; Brown, 2021). To gain legitimacy, they sought support and protection from the rivalrous Holy Roman Empire and the Pope. Thus, in 1311, Milan's feudal overlord, the Holy Roman Emperor Henry VII, made the Visconti clan the *signori* of the city (Bartlett, 2019, 126). Similarly, in 1328, Luigi Gonzaga—the podesta of Mantua—executed a coup in an act of violent treachery against the ruling Bonacolsi family. To legitimate his takeover, he purchased the title of marquis from the emperor, so that his son Ludovico reigned as a prince, and not simply as a *signore* (Bartlett, 2019; Chamberlain, 2020, 141). Such cross-cutting alliances fueled the wars that were ongoing during the midst of the Renaissance (Najemy, 2005, 175–176).

To exercise agency and actively participate in the Renaissance, having military power and legitimacy was insufficient Substantial affluence was also required. Wealth was necessary not only to finance wars and the mercenaries to fight them. Money was also needed to buy off opponents and bribe potential allies, as well as to signify power and influence by means of luxuriant displays of riches and patronage. Fortunately for those involved, Italy was well positioned in this regard. For unlike other European countries, Italy was not entangled in the costly Hundred Years War. Moreover, Italy benefited not only from the most extensive network of commerce and banking in Western Europe but also from a burgeoning industrial base that fueled her subsequent growth and development (Goldthwaite, 2009, 27–30; Ruggiero, 2015, 152–155). Equally important was, according to Goldthwaite (2009):

> … the division of Italy into a multiplicity of states and the existence of a prominent class of merchants and bankers, [so that] wealth was widely distributed, both geographically and socially (G, 31). This process of redistribution, inherent in the economic, political, and social life of Italians, was one of the ongoing dynamics that drove the Italian economy. (31–32)

But, as Najemy makes clear, wealth and military prowess were not enough. Equally important was access to social and cultural capital. As Najemy (2006) describes:

> What made a family great and powerful in the eyes of Florentines was not only is wealth, antiquity, and political offices but also the perception that it commanded the loyalty and support of a greater number of allies, clients, and "lesser" neighbors than did its rivals. (25)

To develop this kind of capital, Renaissance elites competitively engaged in strategic interactions[25] (McClean, 2007). Patronage networks served as the primary channel. Communications within these networks mainly took the form of letter writing, which was highly stylized and modeled according to stringent standards and procedures (McClean, 2007; see also Celenza, 2018, 63–65). These networks were essential for the promotion of the many artistic and cultural endeavors that characterize the Renaissance. But equally—if not more—important, patronage networks played a major role in linking actors together for collaborative purposes, generating over time an emergent social structure. As McClean (2007) explains:

> … each successive effort at networking [took] place from a newly achieved position in the network. Consequently, agency within networks [was] ever adapting to an unfolding social structure as well as an evolving repertoire of discursive gestures. (7)

As detailed here, the history of the Renaissance poses a major paradox: Given the contentious nature of interactions and ongoing warfare, how did the Italian citizenry rise above the conflict so as to generate one of the most vibrant cultures in all of Europe? (Burke, 2014) One answer, of course, is that the struggles for power among Italian elites gave rise to extensive competitive efforts to finance both the arts and the humanities. But that is only part of the story. The other part relates to the nature of collective action. As we have seen in Chapter 2, "How the West Was Won," actors can achieve collective solutions when they share common goals as well as a set of common standards. However, once actors have achieved their joint goals, they will likely contest the outcome, as well as the standards, as each will seek to maximize his or her own benefits at the expense of others (see Chapter 2).

The Renaissance illustrates this dilemma. In the first half of the Renaissance, when independent communes and republics were led for the most part by the *popolo,* who were strong advocates of civic humanism, Italy prospered. The attention of the elites was focused primarily on the challenges of self-government. Subsequently, however, as prosperity continued apace, newly emerging political leaders—the scions of warlords—sought a larger share of the bounty. Hence, they aimed not only to expand their territorial holdings through warfare but also to legitimize their actions and compete with their neighbors based on extravagant displays of wealth as well as support for humanist scholars and artists[26] (Bartlett, 2019; Brown, 2021). The ensuing political conflicts led to the solidification of activities behind territorial boundaries as well as the retrenchment of the humanist standards that had fueled interactions among heterogeneous actors, propelling thereby the Renaissance's emergence. With the hardening of boundaries and the loss of cross-cutting interactions, the Renaissance went into decline.

3.7   Emergent Cities and the Restructuring of Today's Global Economy

The forthcoming chapter on platforms depicts how, in the 13th century, major cities emerged as critical hubs that agglomerated economic actors and greatly expanded trade, reconfiguring the world economy in the process. As well, it describes how the city of Amsterdam—the dominant, and last of these great medieval cities—foundered in the face of rising English economic and military prowess, leading to a phase transition and a reconstructed global economy based on nation-states (Braudel, v. 3 1992; Arrighi and Silver, 1999). This section describes a parallel phenomenon, but one occurring today, which accounts for the emergence of large-scale networked cities that increasingly dominate the world economy.

For an account of the emergence and evolution of today's global cities, we can turn to Allan Scott's book, *A World in Emergence: Cities and Regions in the 21st Century* (2012). Cities, according to Scott, are integral components of capitalism, having arisen together with the emergence of capitalism (Scott, 2012, 2). Explaining the linkages between capitalism and urban spaces, Scott employs the concept of agglomeration (Scott, 2012, 16). Agglomeration, he notes, occurs when the factors of production are assembled together in close proximity, such that their standard interfaces allow them to be easily accessible to those within a shared space. This proximity not only serves to reduce the transaction costs associated with production and exchange; it also allows for the sharing of resources, knowledge, and the administrative responsibilities required for economic interactions. Providing the infrastructure upon which economic activities are agglomerated, cities have served, ever since the 13th century, not only as a gateway for trade but also as magnets for drawing economic actors and resources together in an emergent, self-reinforcing process of cumulative causation. As Scott (2012) affirms:

> Indeed, we can say as a matter of first principles that competitiveness, profitability, and accumulation depend so intimately on agglomeration—among other things—that there is no hitherto realized form of capitalism that is not also associated with urbanization; just as we can say that so far in human history the onward march of capitalist economic development has always, with only minor interruptions, engendered rising levels of urbanization. (17)

Because cities provide a gateway for economic interactions both internally and throughout the larger world economy, the specific form they take and how they are structured is contingent on the stage of capitalism in which they emerge (Scott, 2012). Hence, we see that from the 14th to the 18th century, major cities, which were organized around trading activities, served as the locus of trade. With the rise of manufacturing and the expansion of trade, as well as its extension across the globe, nation-states assumed the mantle, housing within them, and linking together, diverse cities based on standardized Fordist production processes. Given a deepening division of labor, by the 1950s, decentralized industrial facilities had extended well beyond national boundaries to offshore locations (Scott, 2012, 8). Today, the process of reconfiguration is being repeated as economic activities, now centered around "cognitive-cultural" production, have given rise to a new "regionalism" "in which metropolitan areas in many different countries are increasingly caught up in an overarching system of competition, collaboration, and social interaction" (Scott, 2012, 12). As Scott (2012) affirms:

> Large and powerful city-centered regions are now decisively entering on the world stage as important actors in global affairs. Each of these city-regions is composed of an extended urbanized area (often comprising more than one metropolis) and a widely ranging hinterland that may itself contain multiple urban settlements, virtually all of which are assertively inserted in a wider global economy. As such, they are privileged sites of the new cognitive-cultural economy, and this means in turn ... that they are also characterized by many novel features of social organization, international spatial structure, and built form. (14)

3.8 Congressional Chaos and Collapse in the Absence of Standards

Just as standards, and the ways they are organized and bound together, give rise to emergence up the fitness landscape in the cases described above, so we might expect that a lack of coherent standards will lead to dysfunction, dismemberment, and the failure to achieve higher-level outcomes. A prime example of this phenomenon can be found today in the US House of Representatives, where the undoing of its governing rules—that is to say, standards—has led to chaos and disintegration.

Note that, despite their political differences, the Founding Fathers agreed that rules were essential to assure effective governance and democracy (Wolfensberger, 2018). For example, Jefferson (1992) argued that "only by adopting and conforming to such a uniform set of rules of proceeding ... [can] order, decency, and regularity be preserved in a dignified public body" (as cited in Wolfensberger, 2018, 21–22). Likewise, Madison, in the *Federalist # 10*, cautioned that too often, measures are decided "not according to the rules of justice and the rights of the minority party, but by the superior force of an interested and overbearing majority" (as cited in Wolfensberger, 2018, 23). It should come as no surprise then that, given the conflictual nature of our polity today, the focus on rules—that is, who establishes them and by what means—is of great importance (Goldberg, 2018; Mason, 2018). However, in contrast to early American history, when policymakers enacted rulemaking to protect minority rights against an aggressive majority faction,[27] today it is a small minority of House Republicans that are deliberately forestalling collective efforts to govern on behalf of democracy and the national interest (Levitsky and Ziblatt, 2023). As a result, critical policy making has been inhibited, and more recently prevented, by the Rules Committee's Manipulation of House Rules.[28]

Consider, for instance, the drawn-out saga that led to Speaker Kevin McCarthy's ouster and the chaos that followed. Having offended a number of Republican members by compromising with Democrats to avoid a government shutdown, McCarthy was faced with a revolt. Seeking to remain in power, he secretly made more and more compromises—many about rulemaking—with right wing, Republican Members of the House.

Two such compromises were extremely consequential: the first was the appointment of three Tea Party rebels to the powerful Rules Committee; the second was to rule that only one house member was required to instigate an effort to vacate the speaker. McCarthy's efforts backfired profoundly. Unable to gain a majority after fifteen votes and many more compromises, McCarthy was ousted from his position on October 3rd, 2023. Given the bitterness of the ordeal, as well as the new, stringent rule for vacating speakers, the two top contenders for McCarthy's position—Steven Scalise and Jim Jordon—also failed to gain a majority. In a surprise move, Michael Johnson, a new and relatively unknown backbencher was subsequently elected speaker, only to have the rule to vacate hanging over his head. In a surprising and very bold move, Johnson circumvented the rule committee by accepting support from Democratic members to allow a positive vote on a number of foreign aid bills, making his future highly uncertain. However, lacking consensual standards, the Republican House descended into a free-for-all.[29]

Notes

1 As Lyden notes: "… there are no advantages to be gained from treating 'social' as separable from the rules, resources, and the wider structures of power and domination that underpin and legitimate them" (Layder, 2015, 5).
2 The term "social facts" was coined by Emile Durkheim in his book *The Rules of Sociological Method* (1919). The term refers to coercive customs, mores, and norms that constrain individuals in their interactions.
3 Phase transitions refer to sudden changes in a system, reconfiguring its components such that it either collapses or assumes new and greater capabilities. For a detailed discussion, see Chapter 5, "Standards and Phase Transitions in the Middle Ages."
4 Social scientists' preference for the word actor as opposed to the more generic agent reflects the discipline's emphasis on actors having "agency" to control outcomes. As Turner and Machalek point out, "the term 'actor' captures a critical trait commonly attributed to human beings by sociologists, that of "agency," the capacity of being able to act on the world and not be moved passively only by forces that are "external" (environmental) or "internal" (genetic) (Turner and Machalek, 2018, 155–156).
5 For an in-depth discussion of the concept of rules, see Daston (2022).
6 The Cambrian Period occurred approximately 541 million years ago and lasted 54.3 million years. The period witnessed a rapid explosion of new and diverse life forms, including among them vertebrates and the first invertebrates.
7 See Kauffman(2019).
8 See, for instance, Kauffman(2008); see also Kauffman(2006).
9 Arthur(1994).
10 For example, Holland's book, *Adaptation in Natural and Artificial Systems* (1975), which characterized his thinking about learning, evolution, and creativity, and detailed the operation of the genetic algorithm, drew little attention at the time (Waldrop, 1992, 215).
11 Stewart Kaufman developed the concept of "the adjacent possible" to describe the potential possibilities that exist just outside the boundaries of our existing knowledge.
   For an excellent characterization, see Johnson(2010).
12 As described by Wikipedia, autocatalytic sets are "collections[s] of entities, each of which can be created catalytically by other entities within a set, such that, as a whole, the set is able to catalyze its own production."
13 See, for example, Parisi (2023), for the author's comments on scientific communication before the availability of computers.
14 For a description, see: https://arava.org/about-our-community/history-mission/
15 The Golan Heights was a strategic battleground in the Macedonian Wars/Alexander the Great, etc., c. 330 B.C. Most recently, the town of Majdal Shams was attacked by Hezbollah, killing and wounding several dozen young soccer players.
16 It was at this meeting that our student Hyatt asked Peres when Israel would be prepared to return the Golan Heights to Syria. "The time would come," he said, "but when it did, he only hoped that Hyatt would remain in Israel."

17 For a description of the entry of the new director, a Palestinian, see https://open.spotify.com/episode/1n15xyFaOAY5sL0367yyfy?si=751c97cb0dcb4b08&nd=1
18 See Chapter 4 for a characterization of humanism as a platform.
19 For a recent reevaluation of Burckhardt's work, see Bauer and Ditchfield (2022).
20 According to Ruggiero, "… the right to exercise power over others, to rule, was a matter of deep and ongoing uncertainty that significantly colored the way in which governments and other competitors for power presented themselves, claimed authority, and attempted to organize and control society" (Ruggiero, 2015, 55).
21 The guilds were most prominent in the early years of the Renaissance, but they, too, were divided between an upper echelon, which included members of the elite, and a lower order, which was made up of merchants, shopkeepers, artisans, notaries, etc. Well-off guild members were generally more suspicious of lower-class members than they were of the nobility, so they willingly worked with governments comprised of nobles in times of prosperity. But, in times of crises, they supported the lesser members of the *popolo* in their opposition to noble politicians. Guild postures fluctuated significantly as a result (Najemy, 2006, 36).
22 As Najemy claims, "From as early as the twelfth century, elite politics turned on factional conflict, on fierce rivalries between groups of families who did not hesitate to appeal to outside powers for help in their struggles against domestic rivals" (Najemy, 2006, 20).
23 As Ruggiero describes, "The result was messy and violent, but usually seen at the time as pitting the *Popolo* of a city and their new wealth and power based on commercial and artisanal activities against the nobility and their more rural landed wealth" (Ruggiero, 2015, 20).
24 As Najemy explains: "… confrontation served to clarify loyalties, to strengthen the sense of obligation among 'companions and followers,' and thus to sharpen the boundaries between factions" (Najemy, 2006, 26).
25 As McClean asserts: "… each of us constructs a career in the course of our lives constituted by a portfolio of ties to others with whom we are associated. Our careers are made—and we are made—through our interactions with others, as well as through the performance of those tasks to which we have access by virtue of our connections to others. We become more fully the persons we are through interaction, our personhood being constructed out of a number of identities we adopt, singly or in combination, in different interactional settings" (McClean, 2007, 1–2).
26 See, for example, the discussion of Renaissance in Chapter 11.
27 As Wolfensberger points out, upon taking control of the house, new speakers from both parties have typically called for an open rule-making process, only to shift to a closed process that is difficult to navigate in order to get their legislation passed (Wolfensberger, 2018).
28 For a cogent discussion about the House of Representatives rules process, see Rybicki (2022).
29 https://www.washingtonpost.com/politics/2024/04/22/house-republicans-infighting-foreign-aid/

References

Alexander, Jeffrey (1983) *The Modern Reconstruction of Classical Thought: Talcott Parsons*, Berkeley, CA: University of California Press.
Arrighi, Giovanni, and Beverly J. Silver (1999) *Chaos and Governance in the Modern World System*, Minneapolis, MN: University of Minnesota Press.
Arthur, Brian (1994) *Increasing Returns and Path Dependence in the Economy*, Ann Arbor, MI: University of Michigan Press.
Atkinson, Will (2020) *After Bourdieu, A Guide to Relational Phenomenology*, New York, NY: Routledge.
Bartlett, Kenneth (2019) *The Renaissance in Italy: A History*. Indianapolis, IL: Hackett Publishing Company, Inc.
Bauer, Stefan and Simon Ditchfield, eds. (2022) *A Renaissance Reclaimed; Jacob Burckhardt's Civilization of the Renaissance in Italy Reconsidered*, New York, NY: Oxford University Press.
Benson, Robert L. Giles Constable, and Carol D. Lantham (1982) *Renaissance and Renewal in the Twelfth Century*, New York, NY: Harvard University Press.
Berger, Thomas, and Peter Luckmann (1967) *The Construction of Social Reality, A Treatise in the Sociology of Knowledge*, New York, NY: Open Road.
Braudel, Fernand (1992) *Civilization and Capitalism 15th – 18th Century, vol. 3*, Berkeley, CA: University of California Press.

Brown, Alison (2021) *The Renaissance*, 3rd edition, New York, NY: Routledge.

Burke, Peter (2014) *The Italian Renaissance: Culture and Society in Italy*, Cambridge: Polity Press.

Bunge, Mario (2003) *Emergence and Convergence, Qualitative Novelty and the Unity of Knowledge*, Toronto, Canada: University of Toronto Press.

Celenza, Christopher S. (2006) *The Lost Italian Renaissance: Humanists, Historians, and Latin's Legacy*, Baltimore, MD: John Hopkin's Press.

Celenza, Christopher S. (2018) *The Intellectual World of the Italian Renaissance: Language, Philosophy, and the Search for Meaning*, New York, NY: Cambridge University Press.

Chamberlain, Eric Russell (2020) *The World of the Renaissance*, New York, NY: Routledge.

Daston, Lorraine (2022) *Rules: A History of What We Live By*, Princeton, NJ: Princeton University Press.

Demeulenaere, Pierre, ed. (2011) *Analytical Sociology and Social Mechanisms*, New York, NY: Cambridge University Press.

Durkheim, Emile (2014) *The Rules of Sociological Method, and Selected Texts on Sociology and Its Methods*, New York, NY: Free Press.

Elder-Vass, David (2010) *The Causal Power of Social Structures: Emergence, Structure and Agency*, New York, NY: Cambridge University Press.

Garfinkle, H. (1967) *Studies in Ethnomethodology*, Englewood Cliffs, NJ: Prentice Hall.

Goldberg, Jonah (2018) *Suicide of the West: How the Rebirth of Tribalism, Populism, Nationalism, and Identity Politics Is Destroying American Democracy*, New York, NY: Random House.

Goldthwaite, Richard (2009) *The Economy of Renaissance Italy*, Baltimore, MD: Johns Hopkins Press.

Hilgers, Mathieu, and Eric Mangez (2014) *Bourdieu's Theory of Social Fields: Concepts and Applications*, New York, NY: Routledge.

Holland, John (1995) *Hidden Orde: How Adaptation Builds Complexity*, New York, NY: Basic Books.

Jefferson, Thomas (1992) *A Manual of Parliamentary Practice for the Use of the Senate of the United States*, Washington, DC: US Government Printing Office.

Johnson, Stephen (2010) *Where Good Ideas Come From: The Natural History of Innovation*, New York, NY: Riverside Books.

Kauffman, Stuart (1995) *At Home in the Universe, The Search for Self-Organization and Complexity*, New York, NY: Oxford University Press.

Kauffman, Stuart (2006) *Humanity in a Creative Universe*, New York, NY: Oxford University Press.

Kauffman, Stuart (2008) *Reinventing the Sacred: A New View of Science, Reason and Religion*, New York, NY: Perseus Books Group.

Kauffman, Stuart (2019) *A World Beyond Physics: The Emergence and Evolution of Life*, New York, NY: Oxford University Press.

Kuhn, Thomas, and Ian Hacking (2012) *The Structure of Scientific Revolutions*, 50th Anniversary edition, Chicago, IL: Chicago University Press.

Layder, Derek (2015) *Structure, Interaction, and Social Theory*, New York, NY: Routledge.

Levitsky, Steven, and Daniel Ziblatt (2023) *The Tyranny of the Minority*, New York, NY: Crown Book, Random House.

Lewin, Roger (1999) *Complexity: Life at the Edge of Chaos*, Chicago, IL: University of Chicago Press.

Marro, Joaquin (2014) *Physics, Nature and Society, A Guide to Order and Complexity in Our World*, Berlin: Springer.

Martines, Lauro (1979) *Power and Imagination: City-States in Renaissance Italy*, New York, NY: Alfred A. Knopf.

Mason, Lilliana (2018) *Uncivil Agreement*, Chicago, IL: University of Chicago Press.

Maxson, Brian Jeffrey (2013) *The Humanist World of Renaissance Florence*, New York, NY: Cambridge University Press.

Mayfield, John E. (2013) *The Engine of Complexity: Evolution as Computation*, New York, NY: Columbia University Press.

McClean, Paul D. (2007) *The Art of the Network: Strategic Interaction and Patronage in Renaissance Florence*, Durham, NC: Duke University Press.

Najemy, John (2005) *Italy in the Age of the Renaissance*, New York, NY: Oxford University Press.

Najemy, John (2006) *A History of Florence, 1200—1525*, Malden, MA: Blackwell Publishing.

Parisi, Giorgio (2023) *In a Flight of Starlings: The Wonders of Complex Systems*, New York, NY: Penguin Press.

Parsons, Talcott (2013) *The Social System*, New York, NY: Routledge.

Ruggiero, Guido (2015) *The Renaissance in Italy: A Social and Cultural History*, New York, NY: Cambridge University Press.

Rybicki, Elizabeth (2022) *Considering Legislation on the House Floor, Common Practices in Brief*, Washington, DC: US Congress, Congressional Research Service.

Sawyer, Keith R. (2008) *Group Creativity: Music, Theater, Collaboration*, Mahwah, NJ: Lawrence Erlbaum Associates, Publishers.

Sawyer, Keith R. (2005) *Social Emergence: Societies as Complex Systems*, New York, NY: Cambridge University Press.

Scott, Allan (2012) *The World in Emergence: Cities and Regions in the 21ˢᵗ Century*, Cheltenham: Edward Elgar.

Stefan, Bauer and Simon Ditchfield, eds. (2022) *A Renaissance Reclaimed: Jacob Burckhardt's Civilization of the Renaissance in Italy Reconsidered*, Oxford: Oxford University Press.

Sternberg, R. J., ed. (1999) *Handbook of*, New York, NY: Cambridge University Press.

Stones, Rob (2005) *Structuration Theory*, New York, NY: Palgrave Macmillan.

Turner, Jonathan H. and Richard S. Machalek (2018) *The New Evolutionary Sociology: Recent and Revitalized Theoretical and Methodological Approaches*, New York, NY: Routledge.

Waldrop, W. Mitchell (1992) *Complexity: The Emergent Science at the Edge of Order and Chaos*, New York, NY: Penguin.

Wolfensberger, Donald R. (2018) *Congress: From Fail Play to Power Plays*, New York, NY: Columbia University Press.

4

PLATFORMS

Springboards for Evolutionary Outcomes

4.1 The Role of Platforms

Platforms are ubiquitous and take a multitude of forms. Much of the recent attention devoted to platforms highlights their enhanced role in the digital world, especially in the economy (Kenny and Zysman, 2016). However, this emphasis conceals the immense role that platforms have played not only throughout history but also in our evolving universe[1] (Morowitz, 2014; Cusumano et al., 2019). Moreover, much of the literature referencing platforms fails to adequately conceptualize them. Lacking is an understanding of what, despite their diversity, makes platforms unique and what accounts for their ubiquity and persistence over time. Especially important is the question of why platforms loom so large in today's complex environment.[2]

Platforms can best be conceived of as hubs where networks of diverse interactions converge in a multiplexed fashion when actors/actants[3] as well as their purposes and functions align. As such, platforms provide an infrastructure that furnishes time and space for the agglomeration of new resources and diverse information. Drawing on these assets, platform actors generate new standards and practices that give rise to emergent outcomes, allowing them to adapt to changing circumstances. In this sense, platforms are similar to niches; they constitute layers in a hierarchical configuration, providing stepping stones up the fitness landscape.[4]

4.2 Small World Architectures

The characteristic that makes platforms adaptive is their *small world* configuration as determined by the standards that constitute their architectures. (Buchanan, 2002). Bounded and regulated by standards, which—in John Holland's parlance—take the forms of signals,[5] small worlds are characterized by a central core, or cluster, that is made up of strong ties,[6] together with weak ties[7] that extend outward from the core into the environment. Strong ties are essential for facilitating the trust and collaboration required for information exchange and collective action (Coleman, 1988; Krackhardt, 1992; Burt, 1995; Burt, 2007; Borgatti et al., 2009). But these ties, which are subject to homophily, are typically redundant, constraining actors' perspectives and choices. As a result, strong ties tend to inhibit adaptation and reinforce the status quo.

DOI: 10.4324/9781032721125-7

In contrast, weak ties, which spread out beyond a platform, are ideal for retrieving novel, up-to-date information. Moreover, they require fewer resources to maintain. However, these ties require brokers with strong ties to aggregate and integrate new players and information into the platform, so as to generate innovative results (Burt, 1995). When strong and weak ties are joined in a common space, small world platforms not only become highly adaptive; they are, at one and the same time, the source of emergent outcomes (Borgatti et al., 2009). Not surprisingly, small world networks "appear to be pervasive in both nature and human society." As Mark Buchanan (2002) attests:

> ... there is a kind of innate intelligence in these network structures, almost as if they had been finely crafted and laid out by the hand of some divine architect. (21)

When conceived as such, we can see that platforms are complex adaptive systems, in which heterogeneous interactions give rise to new standards that lead to emergent outcomes, better suited to their changing environment. It should be emphasized, however, that emergence is not always positive. Outcomes depend not only on a platform's small world architecture but also on the level and quality of external feedback, as well as on whether and how participants execute the affordances a platform provides. In contrast to mechanical systems, socio-ecological platforms are not in equilibrium; instead, composed of indeterminate actors, they are subject to significant ups and downs. In fact, it is when platforms become static and fail to adapt to their changing circumstances that they collapse in a down-turning phase transition, making way for the emergence of new actors/agents/species, reorganized around new, more adaptive standards, so to reascend the fitness landscape.[8]

4.3 Platforms in Historical Context

To appreciate the recurrent and diverse roles that platforms have played throughout history and, hence, their function in the evolutionary process, we need to focus on the *longue durée,* a time frame characterized by historian Fernand Braudel and the French *Annales* School (Lee, 2012; Braudel, 2014;Burke, 2014). According to Braudel, history should be based not on short-term events, but rather on long-term structures comprised of "ensembles of relationships," Braudel's characterization of long-term structures captures the notion of platforms described above. Below is a description of some very diverse examples.

4.3.1 Governing Platforms: The Regime of Theodoric, King of the Goths, and Ruler of the Romans

In the Barbarian era, powerful tribal leaders emerged within the eastern Roman Empire, where they fought unceasingly to gain supremacy among their peers as well as the approbation and support of the eastern emperor. Maintaining leadership in this tumultuous period required not only continual displays of victory but also the strategic acumen needed to negotiate the complex web of oscillating emperors and shifting barbarian alliances so as to play each off against the other. The stakes were immense, and the outcomes highly uncertain. Because warlords were joined in battle by their families and clans, a staggering military defeat might mean the elimination of an entire people (Geary, 1988; Arnold, 2014; Weimer, 2023).

It was in this context that Theodoric the Great (454–526) emerged as leader of the Ostrogothic people (Geary, 1988; Arnold, 2014; Weimer, 2023). Born to the Amali king Theodemir in the region of Pannonia, Theodoric was taken as a youngster for ransom by Emperor Leo I. It was thus, under Leo's tutelage, that he gained important imperial connections and became well versed in Roman ways. Much like other barbarian leaders, Theodoric was expedient and highly opportunistic, jockeying for power whenever favorable situations arose. In the process, he became a military leader of significant consequence, fighting first with the Huns and then on behalf of the emperor against the warlord Strabo, chief of the Thracian Ostrogoths. However, once this battle was won, the new emperor, Zeno, seeking to neutralize Strabo, gave him Theodoric's position as commander of the eastern Roman forces. Enraged by this betrayal, and needing to sustain his own people, Theodoric retaliated by plundering imperial territories. Failing to vanquish Theodoric, Zeno pursued an alternative approach. To rid himself of his Ostrogothic problem, Zeno ordered Theodoric to make his way to Italy, murder the ruling barbarian king Odovacar, and rule Italy on his own until Zeno might take his place. All of which Theodoric subsequently did, establishing his government in Ravenna, the center of the western empire[9] (Geary, 1988; Arnold, 2014; Weimer, 2023).

On becoming King of Italy, Theodoric successfully adapted his warrior role to one of power broker. To maintain his authority, Theodoric needed to satisfy highly diverse power holders—the Ostrogoth warriors, the Roman senators, and clerical leaders. To grasp the magnitude of Theodoric's challenge, one need only consider the depth and intensity of Italy's divisions (Arnold, 2014, 20–21). Although Romans and Barbarians had increasingly come into contact, their relationship was extremely tense, and their respect for one another practically non-existent. Roman senators prized their "civilized" accomplishments and disdained the Barbarians for the absence of them, whereas Barbarians scorned the Romans for their lack of manliness and military prowess (Geary, 1988; Arnold, 2014, 24–26). Making matters worse, Catholic leaders regarded Theodoric's Arian religion, and that of his followers, as heretical. Yet, for Italy to survive as a viable entity in such an unsettled environment, all players had to stick together. As Ron Burt has argued, in these circumstances, a broker was required (Burt, 1995). To play such a role, Theodoric devised a *governing platform* that made him indispensable not only to the Barbarians and the Roman elite, but also, as importantly, for maintaining the physical and political integrity of Italy (Weimer, 2023).

Instead of trying to unify Italy's divided populace, Theodoric capitalized on its cleavages, keeping the contingents separate from one another and setting himself up as the lone broker who—having a foot in both camps—could integrate their efforts and maintain peace among them. It proved a successful strategy. Theodoric's innovative governing approach remained intact, bringing him renown throughout his thirty-year reign. Characterizing the institutional arrangements supporting Theodoric's regime, Weimer (2023) notes:

In order to place his power in Italy on a firm footing, Theodoric had to transform the Gothic "community of violence" into an ethnically defined class of landowners who performed mandatory military service. His warriors thus assumed a responsibility on behalf of the commonwealth that was intended to make them both indispensable and permanently and conspicuously distinct from the Roman civil administration…. In this way, a kind of dual state came into being: on the one hand, there was a civil administration, which was organized according to the late Roman model. Its personnel consisted exclusively of Romans who spoke Latin…. On the other hand, there was the Gothic military administration. Its personnel were recruited from the army, and it had judicial competence for Goths. The military administration became involved with Romans only when the latter had disputes with Goths. (132–133)

Theodoric's rule was most successful in the area of domestic policy (Weimer, 2023). Notwithstanding ongoing tensions, his Gothic compatriots were well settled in northern Italy, cultivating the land while maintaining their readiness to defend the regime. Likewise, Roman senators and the clerical elite welcomed Theodoric's support for Roman laws, which protected their statuses and guaranteed their privileges. Although Theodoric remained an Arian, he was subdued in his religious posture, lending support not only to Arians but also to Catholicism and the Roman clergy. Most importantly, Theodoric not only reestablished peace and stability in his kingdom; he also reignited aspirations for a return to Rome's imperial past.[10]

Foreign policy presented a greater challenge. Eventually, Theodoric had to address the radical changes in the European environment. For, Theodoric was not the only beneficiary of the dissolution of the western empire. By the end of the 400s, Clovis, leader of the Franks, had established himself in northwestern France; Alaric ruled the Visigoths in Spain; Gundobad reigned in Burgundy; and the Vandals dominated North Africa. With their rise, the balance of power fundamentally changed. Concerned lest these new, power-hungry contenders ignite a war, Theodoric, rather than revising his approach to address this new situation, followed his well-tried script, playing the role of power broker.

To prevent the outbreak of hostilities, Theodoric established marriage alliances with the new rulers, imploring each to pursue peace rather than the risks of war. However, given the increasingly contentious environment, the spoils of war now appeared greater than the risks, so Theodoric's appeals proved unconvincing. Unlike Theodoric's own ascent to power, when options were few and peace and stability were at a premium, the new rulers aimed to make their mark, striking out on their own when opportunities presented themselves. War was the result. This change in circumstances was not, however, the only problem. Theodoric's religious policy likewise suffered a blow. Appearing to favor Arianism over Catholicism, many Romans, as well as the eastern emperor, began to reassess their commitment to Theodoric, especially as other Barbarians converted to Catholicism.

Notwithstanding these trials, Theodoric lived out his reign, but Emperor Julian reassumed control over Italy when Theodoric's offspring proved incapable of enacting a brokerage role (Weimer, 2023). Although Theodoric's governing platform originally matched the needs of both the Romans and the Barbarians, it failed him in the end when he faced a greatly altered set of circumstances and was unable to adapt. Theodoric's problem was structural in part. His bipartite organizational structure spawned strong ties within and firm boundaries around his two sets of constituents. However, in doing so, it greatly limited the development of the weak ties and feedback needed to adapt his political strategy. Theodoric had become so confident of his own success as a power broker that he failed to grasp how fundamentally circumstances had changed. Hence, he was unable to accurately assess the need for a new form of leadership.

4.3.2 Trading Platforms in the Middle Ages: The Champagne Fairs

Trade platforms in the Middle Ages provide a prime example of how networks converge to form platforms of interaction and emergence. By definition, trade requires the convergence of multiple players as well as clearly articulated roles and rules governing their interactions. It was only in Europe's 12th and 13th centuries that these conditions emerged simultaneously, forming webs of interactions that reinforced one another and engendered cumulative causation and the advent of commercial capitalism. As Braudel (1992b) describes:

> Over a very wide area, the crucial move was made from a domestic to a market economy. In other words, the towns were beginning to tower over ... their rural surroundings and to look beyond immediate horizons. This was a 'great leap forward,' in a series that created European society and launched it on its successful career. (94)

Unlike earlier periods, when social, economic, and political changes were incremental and typically local, developments during the central Middle Ages were not only sweeping but also far-reaching, extending well beyond western Europe's boundaries. Among these developments, for example, were significant increases in population, which generated growth in demand as well as in production, especially in the case of agricultural products; the decline of feudalism and the emergence of a money economy, which fostered the rise of cities and towns, along with normative changes that facilitated the division of labor and the commodification of a wide array of goods and services; the Crusades, which expanded trade horizons and new transport routes together with the demand for luxury goods and new credit mechanisms (Abu-Lughod, 1991; Braudel, 1992a, 1992b; Hodgett, 2006; Nicholas, 2006, 71–104; Blockmans and Hoppenbrouwers, 2023, 217–251). Of equal importance for the convergence of interrelated trade networks was the instantiation of innovative institutional arrangements, comprising new standards of interaction, which encouraged investments, reduced risks, and generated the trust and assurances required for trade to flourish (Abu-Lughod, 1991; Hodgett, 2006, 63–64; Nicholas, 2006, 71–104).

Most prominent among Europe's trading platforms were the Champagne Fairs[11] (Abu-Lughod, 1991, 52–77; Braudel, 1992b; Hodgett, 2006, 79–83). As Braudel (1992) relates:

> As for their fame, the popular expression 'not to know your Champagne fairs' meant not to know what everyone else knew. They were a rendezvous for the whole of Europe, for the offerings of both North and South. The trade caravans would converge on Champagne and Brie in assembled and guarded convoys not unlike the other caravans with their camels, which crossed the great deserts of Islam on their way to the Mediterranean. (111)

Emerging as early as 1114 to facilitate trade in local goods between northern and southern Europe, these fairs had, by the late 12th century, widely extended their reach to Italians and northern European merchants who had trading connections in the Levant and further east. More than a single market event, the Champagne Fairs constituted a long-term trading process, which took place sequentially six times a year, with each fair lasting two months and taking place in four different towns located between Champagne and Brie. Once completed, the entire cycle was repeated (Abu-Lughod, 1991, 60–61).

What made the Champagne Fairs unique was their small world architecture. In each participating town, these trade platforms comprised a wide range of diverse actors, linked by both strong and weak ties, who engaged collectively to achieve their joint interest in profiting from trade. Included among these participants were merchants from local towns; local bourgeois servicing out-of-town merchants; long-distance merchants transporting and dealing in high-priced commodities, as well as bankers—most of them Italian—who organized local production, assembled investors, exchanged currency, provided credit, and—at the end of the trading cycle—closed out loans (Abu-Lughod, 1991, 62). Describing this conglomeration of interactions, Chapin notes:

> Each of these groups brought different commodities to the fairs, played different roles in the exchange process, and had a different degree of commitment to the particular fairs in which they participated. The local merchants, hostel keepers, and fair officials of the four towns were, of course, most committed to the Champagne Fairs themselves, for their activities were fully dependent on the prosperity induced by the fairs. Clearly, those local citizens who provided food and lodging to foreigner, who often held their goods and money in safekeeping, and who worked as notaries, local agents, scribes, fair guards, or even porters, had a fully vested interest in the continuation of the emporium.
>
> *(Chapin, 1937, 125–128; Abu-Lughod, 1991, 62–63)*

Notwithstanding these complex interactions, the Champagne Fairs were well-coordinated and highly structured according to multiple standards and regulations. At the top of the platform hierarchy sat the lord, within whose territories the fairs took place. Acting as a broker, he played a proactive role in both establishing these rules and assuring their proper execution (Hodgett, 2006, 63–64). Acting as hosts, lords provided merchants safe passage under their jurisdiction, even indemnifying them for their losses along the way. In addition, they provided the expanse, as well as the infrastructure supporting it, upon which trade took place. Most significantly, the lord established a specific judicial system to resolve disputes and punish rule-breakers. As Abu-Lughod (1991) recounts:

> By the middle of the thirteenth century [the Guards of the Fair] had become a force unto themselves. They had their own seal, different from the count's, recorded summaries of contracts in the Registers of the Fairs, and notarized agreements and enforced their performance …. Their ultimate weapon, however, was to bar from future fairs any grader found guilty of not paying his debts or fulfilling his contracted promises. (59)

Local rulers derived many benefits by brokering the Champagne Fairs. For example, they gained considerable revenues from licensing participants, entrance tolls, fines, and taxes on sales. But even greater were the revenues and indirect benefits derived from the agglomeration and interactions of new, diverse actors, who brought with them innovative knowledge and skills as well as a greater diversity of products and services. Italian bankers, for example, were known not only to provide investment capital but also to encourage and assemble players to engage in local manufacturing (Braudel, 1992b, 112).

Despite the pivotal role that the Champagne Fairs played in fostering a global market economy, they suffered a decline at the end of the 14th century due to their greatly altered operating environment. As Abu-Lughod (1991) explains:

> One of the first lessons to be drawn from the Champagne Fairs is that the external geopolitical factors are absolutely crucial in determining whether or not a locality will be 'strategic' for world trade. It was not the lack of local business acumen that killed the fairs. It was that, in time, the world system outgrew its need for a periodic market in central France. (73)

One negative development was the onset of the plague. Another was the takeover of the fairs by the king of France. In contrast to previous lords, the French King, in amalgamating his territory, established new rules and restrictions, such as tariffs and tolls, that limited the fairs' ability to function as small worlds. But more fundamental changes were also at work. As early as 1200, trade was already moving from the Mediterranean to the North Sea, so that by the mid-14th century, Bruges had become the dominant platform, drawing trade from England to the Baltic. It was here, in Bruges, that the Hanseatic League, a conglomeration of 70–170 northern towns, held together by strong ties and strict rules and regulations, assumed the dominant network platform, linking and brokering the weak ties between their extensive trading partners (Abu-Lughod, 1991, 79–99; Braudel, 1992b, 103; Goldberg, 2012).

4.3.3 Middle Age Cultural Platforms: Renaissance Humanism

Italian humanism emerged in the middle of the 13th century at a time when Italy was undergoing fundamental changes. No longer encumbered by the dominant influences of France and the Holy Roman Empire, Italians were suddenly positioned to strike out on their own. What emerged in this context were a number of city-states and communes, which—as Najemy has described—were "the workshops of politics and government, engines of wealth, and innovative centers of culture, as no European city has been since antiquity" (Najemy, 2004, xxi). Free for the first time to charter their own political paths, these new leaders looked to their Roman past for ideas and inspiration as to how to proceed (Quillen, 2004, 21). As Celenza (2021) details:

> Passion for the Greco-Roman classical world, a love of the ancient Latin language … a love of the humanities, a desire to find 'new' ancient texts (meaning ancient texts that were little— or unknown in the Middle Ages), and a key tendency to compare the current to the ancient world; these became key ingredients in Italian humanism, sustaining and invigorating it. (23)

It was during this search for ancient wisdom that the Italian humanist movement emerged, a movement so encompassing that its impact extended beyond its Italian birthplace and well into the future (Burckhardt, 1995; Greenblatt, 2011). Adherents of humanism engaged in the study of language, or more broadly philology, a field that in premodern times incorporated philosophy. As Celenza attests, because humanists considered poetry, literature, and scholarship a source of human wisdom, they delved into ancient works to gain guidance related to their newfound circumstances (Celenza, 2021, 22–23). As Maxton (2014) maintains:

> From the most basic perspective, humanists sought to inspire moral virtue in their contemporaries by encouraging the study of ethics, the emulation and avoidance of examples from history—and to a lesser extent literature—and a firm knowledge of the grammatical and empirical tools necessary to move other to their opinion of the morally correct point of view. (10–11)

One of the early, and most influential, pursuers of ancient texts was Petrarch, a gifted poet and scholar (1304–1374), who, having been inspired by the works of Cicero, implored his cohorts not only to collect and analyze works of antiquity but also to write and debate one another in classical Latin, a language that for Petrarch, and many others, epitomized beauty as well as fundamental truth (Greenblatt, 2011, 116–117). As importantly, in contrast to the passive life characteristic of the Medieval period, these humanists called on individuals to become active participants in society (Burckhardt, 1995). Indeed, language, they argued, provided the means of doing so. As Quillen (2004) points out:

> By arguing for a connection between 'eloquence,' which [Petrarch] took to mean clarity and persuasive force as learned from ancient authors, and 'virtue' defined as the capacity to live a fully human life, he gave humanism a compelling rationale that transformed it into a movement. (24)

Petrarch's advocacy for humanism gained momentum in his native Florence, where it inspired the poet Giovanni Boccaccio and Coluccio Salutati, the Chancellor of the Florentine Republic, to promote it. Indeed, it was under Salutati's leadership that humanism blossomed, as he employed his position and authority to garner support for humanism and to recruit adherents to the philosophical and civic ideas embodied within it (Celenza, 2018, 59). Lacking a strong scholastic tradition, as well as a dominant ruling elite, Florence was singularly receptive to humanism, where it appealed to the ruling populo, who built upon its teaching to justify its participatory governing philosophy (Najemy, 2004; Quillen, 2004, 26–27; Celenza, 2018, 59–63). By the 15th century, as more and more patricians came to believe that a classical education was the best preparation not only for a well-grounded citizenry but also for success in life, humanism came to dominate the city's culture, and from there it spread out to Naples, Venice, Milan, and Rome (Quillen, 2004, 29–31). As Maxson (2014) describes:

> Increasingly over the fifteenth century, Florentines presented themselves as versed in the classics in order to maintain and earn capital for themselves, their families, and their city. By the mid-fifteenth century at the latest, large numbers of Florentines had been educated according to humanist-style curricula, were hiring humanist tutors for their children, were learning and imitating classical rhetorical techniques, were copying and reading humanist texts, and were commissioning works of cultural production inspired by their studies. As many as two-thirds of Florentines in 1427 were literate, at least in the vernacular. (17)

While many accounts of Italian humanism focus on individual participants and texts, humanism was, in fact, a collective effort, greater than the sum of its parts, which is to say, it was *a movement* that served as a *platform* for the social and cultural changes associated with the Renaissance (Revest, 2013, 427). How, one might ask, did such a collaborative effort come about and become transformed into a movement? To answer this question, consider the insights of Randall Collins. According to Collins, individuals coalesce and form identifiable groups through their symbolic interaction chains, which generate emotional entrainment and experiences that serve as magnets of cultural significance, where "cultural is created, denigrated, or reinforced" (Collins, 2014, 13). During the Renaissance, highly standardized modes of letter correspondence, distributed among humanist cohorts, formed the primary mode of interaction.[12] As McLean (2007) points out:

> Patronage letter writing was an institution in Florentine society: regular sequences of activity that supported and reproduced a set of shared expectations about the world and how it operates. It was an important tool for trying to achieve social mobility, security, and the recognition of others. It was, in short, a critical part of Florentine culture. (xiv)

And so it was conceived by its participants. For humanists believed they were taking part in "one far-reaching movement which, by renovating language, was truly renovating civilization itself; a belief that they were witnesses and actors of a glorious era, ushered in by already famous pioneers and by a series of acquisitions … leading to the imminent consecration of what we now call humanism" (Revest, 2013, 430–431).

What made humanism so prolific and enduring was its small world architecture. On the one hand, it was comprised of a core of dedicated participants whose primary activity was in finding, translating, and writing classical Latin texts. These people, who were relatively few in number, were connected by very strong ties. Linked to this core of humanists was a much larger group of—one might say—amateur humanists, who—connected by weak ties—served as the primary audience of, and distributors for, the core of humanists. The most predominant of these were the social elite, who had the social status and financial wherewithal to propagate humanism (Maxson, 2014, 13–14).

4.4   Accounting for the Rise of Platforms Today

To illustrate the ubiquity of platforms, the preceding discussion focused on the role of platforms in very diverse, historical contexts. In this section, I consider the rise of platforms today, as they have evolved in a rapidly expanding global context. As argued above, platforms are a ubiquitous phenomenon throughout the universe, providing stepping stones to higher levels of fitness. Thus, as we might expect, in today's world, in which interactions have expanded and become more complex, the need for platforms is all the greater.

4.4.1   Business Platforms—the Case of IBM

The history of IBM is legendary, not only because of the company's longevity but also, and perhaps more telling because of its ability, in times of crisis, to successfully adapt to its radically altered circumstances. As Cortada (2023) describes:

> IBM's ability to change, while not always as effective as one might want, was sufficient to sustain the firm …. Each transition took at least a decade to accomplish. In the process, careers were made and broken, and new product lines were created and discarded, while customers encouraged the firm to change. Each of these changes were made when senior executives saw the need for change. Furthermore, each transition affected the next one. (27–28)

A key factor in IBM's ability to adapt to its changing environment was its creation of platforms, both for its products and its business, which functioned as small worlds, tightly linking employees and customers together. Equally important was IBM's establishment of an enduring business culture that actively promoted and supported its small world environment.

Born into an era of tremendous prosperity, in which data analysis was in great demand, IBM was able to reap the benefit of these circumstances by generating a core business dedicated to tabulating cards. It proved to be a highly successful business strategy. By leasing its equipment to its customers while selling the cards required for their use, IBM not only generated an increasing number of reliable sales but also—like in a small world—locked its customers into IBM (Cortada, 2023, 67).

In fact, tying customers to IBM through sales and outreach was fundamental to the company's long-term *modis operandi*. It did so not only through its technology policy but also, and as importantly, by building strong, external relationships. To this end, members of IBM's salesforce engaged intensely with its customers. Rather than developing products and services from the top down, IBM typically responded to customers suggestions and needs. As well, to promote loyal customers, salespersons went so far as to train them in using equipment and to conduct financial analyses justifying a technology's purpose and use. As Cortada (2023) describes, it was by virtue of these interactions that:

> IBM became the hub, the center of the tabulating data processing ecosystem, the go-to place for goods and information, and later, management practices, for organizations automating their largest information processing activities. When computers came along, a new generation of IBM executives understood the need to hop onto the new technology that otherwise would replace their tabulating machines. (88–89)

As sales increased and branch offices were established, IBM called on its employees to actively generate and consolidate interactive ties. Salespersons, for example, were required to live in the same locales as their customers and to actively engage in their communities' social and economic activities. Thus, they joined the same country clubs, participated in local civic organizations, attended the same churches, and became members of the social and political elite (Cortada, 2023, 52–53). Policies such as these were especially important as IBM expanded into Europe, where the board of directors were comprised of local elites, and the branch offices were staffed by native citizens trained in the IBM culture (Cortada, 2023, 111).

To balance these outreach efforts, IBM consciously developed a strong inner core. To this end, it established, and continually reinforced, a foundational business culture, which it reinforced through pervasive symbols and rituals (Cortada, 2023, 48–49). Much as Randall Collins has attested, groups are solidified through the reinteraction and reenactment of stories, memes, and rituals, which not only promote a common identity but also, and as significantly, heighten a member's commitment to achieving the group's goals (Collins, 2014). At IBM, such efforts proved highly successful. As Cortada (2023) relates:

> By the mid 1920s, there had emerged an information ecosystem involving employees and customers and an effective corporate culture supporting that community. The company's culture facilitated the functioning of the ecosystem, with many activities supporting this community, reflecting company values and desired behaviors. (72)

A critical feature of IBM's culture was its focus on sales. To imbibe the IBM culture among the company's many employees, Thomas Watson established a "Sales School," where the company's values, selling techniques, and operational procedures and products were inculcated. As importantly, in factory sites, customers participated in the same training, so that "everyone shared common beliefs and practices [allowing] individuals [to] make decisions and take actions confident that they were doing the right thing" (Cortada, 2023, 39). What emerged from these joint efforts was an information ecosystem involving both employees and customers based on an effective corporate culture that nurtured and reinforced the IBM business community (Cortada, 2023, 71).

Grounded in its successful business strategy and supported by a corresponding business culture, IBM reached its peak performance at the end of the Second World War. Not only had the company benefited from a greatly expanded global market; its profits and earnings were higher than ever. Moreover, IBM's engagement in the war effort had led to greater innovative efforts, especially in electronics, thereby preparing IBM to expand its product line into computing (Cortada, 2023, 130). However, as we have seen, it is when systems are at their height and bound together by tight interconnections that they are least able to adapt to changing circumstances. How, then, we might ask, was IBM—given its well-established success and entrenched business culture—able to transform itself from a relatively low-tech card tabulating company into one focused on electronics and computing?

As a result of its wartime experience, IBM enjoyed many advantages moving forward. Much as the company had instantiated itself as a card tabulating company at a time when data processing was in great demand, so too, in the post-war period, it entered the computer business at a point when computing was on the rise (Cortada, 2023, 175). Nevertheless, making the transitions from tabulating cards to computers entailed a number of risks and challenges. As Cortada (2023) notes:

> Increase by at least one order of magnitude the challenges IBM faced compared to those encountered by Watson Sr.'s generation, increase the number of employees by two orders of magnitude, and more than double the number of countries where IBM did business, and we begin to sense the magnitude of the tectonic events IBM faced. Size, success, and influence made IBM well known, but the status carried with it more than revenue and prestige; it included grave risks and consequences. (150)

In addressing these challenges, Thomas Watson was well aware that, to successfully compete in this new environment, the company needed to upgrade its knowledge in electronics and computing. To do so, he extended the company's reach well beyond the traditional bounds of IBM to those places—such as universities, the military, and other businesses—where cutting-edge computing innovations were taking place. Of equal importance was Watson's understanding that to make the most of new technologies required reconfiguring IBM itself, so as to meet the changing structure of demand (Cortada, 2023, 182–183). However, with Watson's passing, it was left to Watson's son, Thomas Jr., who was less embedded in IBM's traditional culture, to effectively execute these ideas.

On taking over his father's role, Tom Jr. set about reorganizing IBM. He knew that, to reduce its bureaucracy and thereby tame the problems associated with IBM's extensive growth and complexity, a major restructuring was called for (Cortada, 2023, 190). Thus, at an early management meeting, he announced his plan to replace his father's centralized leadership approach with one that called for distributed responsibility and decision-making authority throughout the company (Cortada, 2023, 190). Commenting on the broad sweep of Thomas's plan, one executive opined, "In three days we transformed IBM so completely that almost nobody left that meeting with the same job he had when he arrived" (as cited in Cortada, 2023, 191). Not surprisingly, it was not long thereafter that the old-timers at IBM were gone, and new leaders emerged, governing now over a number of department platforms, most of whom were, on average, only forty years old.

Notwithstanding these far-reaching changes in its organizational structure, IBM continued to carry out its business in keeping with its long-established corporate culture linking employees together with its customers. A prime example of this effort was IBM's involvement in the user group Share. Because computing was a fast-growing emerging business, IBM engineers and users of its technology were on much the same learning curve. To gain insights about developing computing technologies, IBM customers organized themselves into a user group where they shared information regarding the full range of IBM products. Although Share was run by volunteers, IBM salespersons and engineers were active participants, both providing information and learning from their customers about their needs and concerns (Cortada, 2023, 179–181). The pay-off proved considerable when IBM set out in the early Sixties to generate the next level of computer technology.

Although IBM has traditionally paced its innovations in response to its customers' needs, in the early Sixties, the company took a great leap forward, and indeed an incredible risk, in developing its transformative, but at times controversial, System 360. Developers of the system faced major challenges. To begin with, they had to design compatible, plug-in hardware system components that surpassed the performance of existing technologies. Entailed in this effort was the development of computer chips, an expensive enterprise still in its infancy. As compelling, to ensure compatibility and system interoperability required an entirely new and extremely complex undertaking that entailed developing millions of lines of computer code. However, after four years in development, IBM was able, in April 1995, to offer a system of six advanced computers, together with a total of 150 peripheral products to its customers. Most noteworthy was the development of common standards that established interoperability throughout the system, allowing IBM to establish a small world platform that locked in its customers for more than two decades (Cortada, 2023, 211–217).

In developing System 360, technology was not the only challenge. Equally problematic was the task of rallying the requisite, diverse talents from across different company sectors and—despite their divergent perspectives—coordinating their efforts. As one might expect, disagreements were commonplace, and the situation was at times chaotic (Cortada, 2023, 205). But keep in mind, much as we have seen in other complex cases, conflict among actors can be a critical source of innovation. And so, it was at IBM. When confronted with a rivalry between two key engineering divisions, Vin Learson—the vice president of manufacturing and development—employed his problem-solving talents to apply "abrasive interaction," a management technique that required disputants to stand in each other's shoes. As both sides began to collaborate in developing System 360, a compilation of new technologies emerged, which quietly shaped computing events around the world' (Cortada, 2023, 209–211, 233).

Decades later, IBM was far less successful in adapting to a radically new set of circumstances, brought on by the rise of personal computers. Five years after the PC's introduction in 1981, it was becoming commodified, such that, for the first time, IBM was facing intense competition from a variety of new providers such as Compaq and Dell. As Cortado (2023) describes:

> From IBM's perspective, in the first half of the 1980s, something terrible began to happen, turning a new and vast market into an ugly battleground with many rivals. From the second half of the 1980s until 20 years later, the PC story degenerated into a disappointing chapter in IBM history. IBM's customers became unnerved, the company stumbled badly, and the PC business became the graveyard for the careers of a slew of employees and potential future CEOs. Over time, it became clear that the disappointing story of IBMs PC mirrored the declining performance of the entire company. (379)

Why did IBM fail to adapt in the 1980s when it did so successfully in the 1960s? As argued below, although IMB made a number of mistakes in developing the PC—such as losing control of its operating system and failing, when it had the opportunity, to partner with Microsoft—the overriding cause of IBM's decline stemmed from a complexity catastrophe brought about by its increasingly immutable, bureaucratic structure, which greatly inhibited IBM's ability not only to improvise in response to changing circumstances, but also to maintain its dominant business platform. Given its phenomenal growth throughout the Seventies, IBM became exceedingly bureaucratic as it sought to manage its numerous divisions and the ongoing turf battles taking place among them. Overall governance of these divisions was executed by a senior management council, chaired by the CEO. Unfamiliar and unattuned to the rapidly advancing computer environment, the council's members, lacking a visionary perspective, stifled innovation and generated "micromanaging contradictions, rivalries, and competing perspectives" (Cortada, 2023, 416).

It was within this context that IBM's PC both rose and declined. Unlike IBM's other technologies, the PC's development occurred outside the purview of IBM's formal, bureaucratic processes. With the approval of Frank Cary, IBM's CEO, shepherding the PC's development was undertaken by Don Estridge, a nonconformist engineering manager known for skirting company norms (Cortada, 2023, 388). Sheltered from the bureaucracy, Eldridge and his engineering cohorts brought the innovative PC to fruition. However, once developed, IBM management incorporated the PC into the IBM bureaucracy, where rivals in the firm successfully reined it in. Estridge was replaced by Bill Lowe, a member of IBM's "old school," who had far less PC experience and even less foresight about the emerging field of computer networking. With this move, IBM's platform collapsed, as the perennial link between IBM and its customers was cast asunder. It took years for Big Blue to recover.

4.4.2 Today's Digital Platforms

Today, the business world has become increasingly focused on the emergence of digital platforms and the opportunities they present for enhanced economic performance. Thus, platforms are touted not only for their ability to generate economic growth and innovation, as well as the efficiency of the economy, but also to reduce transaction costs, encourage new forms of collaboration, facilitate users' access to services, and enhance businesses' competitiveness by locking in customers and users (Gawer, 2009; Parker et al., 2016; Cusumano et al., 2019). The phenomenal success of Amazon, Facebook, Google, and Apple—the top four digital platforms—confirms these positive assertions (Moore and Tambini, 2021). In achieving their phenomenal gains, these four companies have assumed control over much of the online space (Moore and Tambini, 2021; Van Dijck et al., 2018).

Successes such as these have not only generated acclaim from economists and business pundits; they have also brought into relief concerns about the societal impacts of digital platforms (Van Dijck et al., 2018; Jorgensen, 2019). Some critics have argued, for example, that business analysts have scoped their analysis of digital platforms far too narrowly within a neoclassical framework (Mansell, 2019). They contend that these analysts, by focusing solely on price and consumer welfare, skirt the full range of their impacts, such as those experienced by citizens, communities, public values, the media, etc. (Van Dijck et al., 2018). Other concerns about digital platforms relate to the potential social, economic, and political dangers arising from the monopoly power associated with digital platforms today (Wu, 2018).

In sorting out these developments, we need to consider that, although the rise of digital platforms is a new phenomenon, platforms have, themselves, been a universal constant. As detailed above, platforms constitute an essential component of the social order. However, unlike today's digital platforms, those of yesterday were taken for granted, for the most part. Why was this the case? Addressing this question allows us not only to better understand what makes digital platforms distinct but also what—given their unique attributes—might be the best way to govern them. As postulated below, the answers to these questions relate to both the nature of digital technologies themselves, as well as to the radically different set of circumstances in which these digital platforms have emerged.

The proliferation of platforms today is a function of the ongoing complexity of the universe. At each level of the universal hierarchy, platforms agglomerate actors and resources, as well as the standards driving their interactions, so as to generate emergence and ascent up the fitness landscape. Embodying greater resources and information, emergent systems engender higher levels of complexity, requiring the assembly of ever more numerous, and diverse, platforms, serving ultimately as the infrastructure for evolution.[13] In contrast to earlier, less complex, societies, when platforms were loosely coupled and limited in scope, today's platforms, which comprise platforms within platforms, are not only far greater in scope; they are also densely integrated in complex ways, based on digital technologies and the algorithms governing them (Ezrachi and Stucke, 2016). As telling, unlike earlier platforms, which were embedded in and governed by larger social, political, or economic environments, today's platforms are increasingly colonizing their own environments, such that they are dis-embedded from and no longer accountable to them. Absent a dynamic, external environment, we can expect the digital platforms of the future to function less and less as small worlds. Based on our understanding of complexity, we might suppose that digital platforms will become enveloped at some point in a complexity catastrophe, such that, over the *longue durée,* they will collapse in a phase transition.

Notes

1 Platforms can be viewed as 'social structures' where interactions not only take place but also have agency in their own right. It is by virtue of their agency that emergence takes place. For one discussion, see Layder (1981).

2 Ametowobla, Dzifa, and Kirchner (2023).

3 The word "actant" is derived from Actor Network Theory. It refers to technologies that have agency of their own. See Bruno Latour (2007).

4 Kauffman (1995, 9–10).

5 While Holland (2014) has employed the ecological metaphor *niche* to represent system structures, I employ the more generic term "platform" to depict the locus of multiplexed relationships that, by virtue of their interactions, give rise to complex, emergent outcomes. Viewing structures from a socio-ecological perspective, I prefer the term platform because, while enveloping the idea of niches, it can also be applied across the many contexts in which emergence takes place. In addition, whereas Holland characterizes the mechanisms giving rise to complex, emergent outcomes as signals, I classify signals as an instance of standards, which are defined more generally as interfaces governing interactions, as described in Chapter 1.

6 Dense ties occur when individuals are interconnected by more than one feature, such as gender, class, religion, place of origin, profession, etc. While dense ties foster trust and collaboration, they tend to be risk-averse and reluctant to change. For discussions, see Krackhardt, D. (1992) and McPherson et al. (2001).

7 According to sociologist Mark Granovetter, *weak ties* are more beneficial than strong ties in gathering information because they are less likely to be redundant. See Granovetter, Mark (1973).

8 According to Stuart Kauffman, such ups and downs are a signature of complex adaptive systems. Illustrating this process, Kauffman points to the Cambrian period of evolution. As he describes, "… recent evidence suggests that the highest rate of extinction, as well as speciation, [occurred during] the Cambrian itself. Over the next million years, the average diversity of species increased to a kind of rough steady state. But that level was, and persistently is, perturbed by small and large avalanches of extinction that wipe out modest numbers or large numbers of species, genera, or families.… speciation and extinction seem very likely to reflect the spontaneous dynamics of a community of species. The very struggle to survive, to adapt to the small and large changes of one's coevolutionary partners, may ultimately drive some species to extinction while creating novel niches for others" (Kauffman, 1995, 28–29).

9 The fifth century witnessed the disintegration of the Western Roman Empire, as military leaders—a number of whom were non-Romans—came to overshadow civilian leaders and assume greater and greater power and control. Thus it was that, in 476, the barbarian leader Odovacar wrested power from the Western Emperor Orestes, dethroned his young son, Romulus Augustus, and declared himself king of Italy. Odovacar's subsequent military exploits in the east threatened the emperor Zeno, leading him to send Theodoric west to eliminate Odovacar and assume control of Italy. In 493, after three years of battle, Theodoric—in an act of treachery—slayed Odovacar by his own hand. For a detailed account, see Weimer (2023, 98–129).

10 As Arnold describes, "This conscious appeal to the late republic and early principate made it possible for Italo-Romans to laud Theodoric as a new Trajan, a new *optimums princeps* who often imitated the first. It also helped to transform Italy from the decadent Roman Empire of the fifth century into the glorious "republic" of the first century, a period worthy of admiration and imitation in those apparently trying times. Legitimacy was thus gained for Theodoric among Italo-Romans through his princely appellation and its ideological trappings; nor was he the first late antique ruler to understand their power within a specifically Italian context (Arnold, 2014, 87–88).

11 Commercial fairs emerged early and were prevalent in the central Middle Ages. As Abu-Lughod points out, in the post-feudal era, "small trading centers grew up or were eventually revived around the castles of the warlords or within the protected radii of monasteries, where goods could be exchanged by merchants who obtained special dispensations and protection from the local lords in return for their commercial services. Some of these settlements, particularly those located where overland or river routes crossed, eventually served as sites for periodic fairs or as exchanges for the minuscule amount of 'international trade' that reached the interior"(Abu-Lughod, 1991, 44–45).

12 According to McLean, "Recipients of cultural signals try to read identities off of actors" embeddedness in social structures, but senders also actively represent that embeddedness in their messages. … And each successive effort at networking takes place from a newly achieved position in a network. Consequently, agency within networks is ever adapting to an unfolding social structure as well as an evolving repertoire of discursive gestures (McLean, 2007, 7–8).

13 A parallel development occurred during industrialization, which witnessed greater specialization and an increase in the division of labor.

References

Abu-Lughod, Janet L. (1991) *Before European Hegemony: The World System, A.D. 1250–1350*, New York, NY: Oxford University Press.

Ametowobla, Dzifa, and Stefan Kirchner (2023) "The Organization of Digital Platforms—The Role of Digital Technology and Architecture for Social Order," *Zeitschrift fur Soziologie* 52 (2), 1–13.

Arnold, Jonathan J (2014) *Theodoric and the Roman Imperial Restoration*, New York, NY: Cambridge University Press.

Blockmans, W., and Hoppenbrouwers, P. (2014) *Introduction to Medieval Europe 300-1500* (2nd ed.). London: Routledge.

Borgatti, Stephen P., Ajay Mehra, Daniel J. Brass, and Giuseppe Labianca (2009) "Network Analysis in the Social Sciences," *Science* 323 (892), 1–5.

Braudel, Fernand (1992a) *The Wheels of Commerce, v.2, Civilization and Capitalism 15th-18th Century*, Berkeley, CA: University of California Press.

Braudel, Fernand (1992b) *The Perspective of the World, v.3, Civilization and Capitalism 15th-18th Century*, Berkeley, CA: University of California Press.

Braudel, Fernand (2014) *On History*, Chicago, IL: University of Chicago Press.

Bruno, Latour (2007) *Reassembling the Social: An Introduction to Actor-Network Theory*, New York, NY: Oxford University Press.

Buchanan, Mark (2002) *Nexus: Small Worlds and the Groundbreaking Science of Networks*, New York, NY: W.W. Norton and Co.

Burke, Peter (2014) *The Italian Renaissance: Culture and Society in Italy*, Cambridge: Polity Press.

Burckhardt, Jacob (1995) *The Civilization of the Renaissance in Italy*, New York, NY: Random House, Modern Library Edition.

Burt, Ron (1995) *Structural Holes: The Social Structure of Competition*, New York, NY: Oxford University Press.

Burt, Ron (2007) *Brokerage and Closure: An Introduction to Social Capital*, New York, NY: Oxford University Press.

Celenza, Christopher S. (2018) *The Intellectual World of the Italian Renaissance: Language, Philosophy, and the Search for Meaning*, New York, NY: Cambridge University Press.

Celenza, Christopher S. (2021) *The Italian Renaissance and the Origins of the Modern Humanities: An Intellectual History*, New York, NY: Cambridge University Press.

Chapin, Elizabeth (1937) *Les villes de foires de Champagne des origins au debut du XIVe. Siècle*, Paris: H Champion.

Coleman, James. S. (1988) "Social Capital and the Creation of Human Capital," *American Journal of Sociology* 94, S95–S120.

Collins, Randall (2014) *Interaction Ritual Chains*, Princeton, NJ: Princeton University Press.

Cortada, James. W. (2023) *Inside IBM: Lessons of a Corporate Culture in Action*, Columbia Business School Publishing.

Cusumano, Michael A., Annabelle Gawer, and David B. Yoffie (2019) *The Business of Platforms: Strategy in the Age of Digital Competition, Innovation, and Power*, New York, NY: Harper Collins.

Ezrachi, Ariel, and Maurice E. Stucke (2016) *Virtual Competition: The Promise and Perils of the Algorithm-Driven Economy*, Cambridge MA: Harvard University Press.

Gawer, Annabelle, eds. (2009) *Platforms, Markets, and Innovation*, Cheltenham: Edward Elgar.

Geary, Patrick J. (1988) *Before France and Germany: The Creation and Transformation of the Merovingian World*, New York, NY: Oxford University Press.

Goldberg, Jessica L. (2012) *Trade and Institutions in the Medieval Mediterranean*, New York, NY: Cambridge University Press.

Granovetter, Mark (1973) "The Strength of Weak Ties," *The American Journal of Sociology* 78 (6), 1360–1380.

Greenblatt, Stephen (2011) *Swerve: How the World Became Modern*, New York, NY: W. W. Norton and Company.

Hodgett, Gerald A. (2006) *A Social and Economic History of Medieval Europe*, New York, NY: Routledge.

Holland, John (2014) *Signals and Boundaries: Building Block for Complex Adaptive*, Systems, Cambridge, MA: MIT Press.

Jorgensen, Rikke Frank, ed. (2019) *Human Rights in the Age of Platforms*, Cambridge, MA: MIT Press.

Kauffman, Stuart (1995) *At Home in the Universe: The Search for Laws of Self-Organization and Complexity*, New York, NY: Oxford University Press.

Kenny, Martin, and John Zysman (2016) "The Rise of the Platform Economy," *Issues in Science and Technology*, National Academies of Science, Arizona State University, 61–69.

Krackhardt, D. (1992) "The Strength of Strong Ties: The Importance of Philos in Organizations,". in D. Nohria and R. Eccles, eds. *Networks and Organizations: Structure, Forms, and Actions*, Boston, MA: Harvard Business School Press.

Layder, Derek (1981) *Structure, Interaction, and Social Theory*, London: Routledge & Kegan Paul Ltd.

Lee, Richard E. (2012) *The Longue Durée. And World-Systems Analysis*, New York, NY: State University of New York.

Mansell, Robin (2019) "Bits of Power: Struggling for Control of Information Networks," *The Political Economic of Communication* 5 (1), 2–29.

Maxson, Brian Jeffrey (2014) *The Humanist World of Renaissance Florence*, New York, NY: Cambridge University Press.

McLean, Paul D. (2007) *The Art of the Network: Strategic Interaction and Patronage in Renaissance Florence*, Durham, NC: Duke University Press.

McPherson, Miller, Lynn Smith-Lovin, and James M. Cook (2001) "Birds of a Feather: Homophily in Social Networks," *Annual Review of Sociology* 27(1), 415–444.

Moore, Martin and, Damian Tambini (2021) *Regulating Big Tech: Policy Responses to Digital Dominance*, New York, NY: Oxford University Press.

Morowitz, Harold J. (2014) *The Emergence of Everything: How the World Became Complex*, New York, NY: Oxford University Press.

Najemy, John M. (2004) *Italy in the Age of the Renaissance: 1300–1500*, New York, NY: Oxford University Press.

Nicholas, David (2006) "Economy," in Power, Daniel, ed. *The Central Middle Ages*, New York, NY: Oxford University Press.

Parker, Geoffrey G., Marshall W. Van Alstyne, and Sangeet Paul Choudary (2016) *Platform Revolution: How Networked Markets Are Transforming the Economy and How to Make Them Work for You*, New York, NY: W.W. Norton & Company.

Quillen, Carol Everhart (2004) "Humanism and the Lure of Antiquity," in Najemy, John M., ed. *Italy in the Age of the Renaissance 1300–1500*, New York, NY: Oxford University Press.

Revest, Clemence (2013) "The Birth of the Humanist Movement at the Turn of the Fifteenth Century," *Annales HSS* 68 (3), 425–456.

Van Dijck, Jose, Thomas Poell, and De Waal (2018) *The Platform Society: Public Values in a Connected World*, New York, NY: Oxford University Press.

Weimer, Hans-Ulrich (2023) *Theodoric the Great, King of the Goths, Ruler of Romans*, New Haven, CT: Yale University Press.

Wu, Tim (2018) *The Curse of Bigness: Antitrust in the New Guilded Age*, New York, NY: Columbia Global Reports.

5

STANDARDS AND PHASE TRANSITIONS IN THE MIDDLE AGES

5.1 Introduction

This chapter examines the evolution of the Middle Ages (1000–1500) in the context of complexity theory. It focuses on how changing configurations of standard interactions engender major phase transitions. Whereas most historical analyses are centered on a sequence of singular, short-term events leading to changes in the social order, I follow Fernand Braudel[1] in pursuing the longue durée, which characterizes history as being based, not on short-term events, but rather on long-term, interlinked, enduring phenomena[2] (Burke, 1990, 35–62). Tracing societal movements up and down the fitness landscape,[3] I center my inquiry, much like Braudel, on "ensembles of changing relations" (or, as I characterize them, changing standard configurations), a key explanatory variable in complexity theory. When these changes occur on a major scale, I conceive of them as phase transitions.

Phase transitions refer to a sudden change in the state of a property that leads to its physical reorganization and revised capabilities. Describing these phenomena, physicist Per Bak notes that there is a tendency in large-scale complex systems to evolve into poised "critical states"[4] in which minor disturbances and dynamic interactions trigger monumental events, or phase transitions (Bak, 1996, 12–13; Sole, 2011). As an example, phase transitions from water to ice result not from the alteration of the water molecules themselves but rather from a shift in the overall organization of the molecules that make up water. Moreover, Buchanan emphasizes that phase transitions are not limited to the material world. As the fall of Rome and the rise of Napoleon clearly demonstrate, such changes propagate throughout human society, due to the changing fabric of human relations, whether cultural, sociological, economic, technological, or political[5] (Buchanan, 2002, 237; Sawyer, 2005).

5.2 Focusing on the Fitness Landscape

Changed states and reconfigured relationships open the way for the emergence of new types of interactions and social structures higher up on the fitness landscape (Kauffman, 1995). Such was the case during Europe's long Middle Ages when Western Europe witnessed a great

DOI: 10.4324/9781032721125-8

expansion together with the consolidation of regional territories behind definitive state boundaries (Blanning, 2003; Pennington, 2014, 47–48). As feudalism receded, the vast empires, such as the Hapsburgs, which had been organized based on family dynasties cutting across territorial boundaries, were subsequently overtaken by coherent, autocratic states (Curtis, 2013). The result was an increase in fitness—that is to say, competence in relationship to a changing environment—together with the total reconfiguration of the map of Europe.

To account for this historical restructuring, this chapter probes key phase transitions throughout the Middle Ages that transformed Western Europe's developmental path. To keep my account manageable, I limit my analysis to key societal transitions, although I recognize that disasters—such as famines and health crises like the plague—were contributing factors as well as phase transitions in their own right. Moreover, I identify the existence of phase transitions based on the formation and collapse of interactive ties. Employing this approach, we see an overriding pattern whereby actors achieve the height of their power and performance when ties are densely entangled—as Kauffman says, at the edge of chaos—only to fall back, in a complexity catastrophe,[6] to broken ties and semi-isolated states when circumstances change and chaos takes hold (Kauffman, 1995, 28–29). As complexity theory posits, rebuilding then, with rearranged parts, leads to adaptation and higher levels of fitness, as measured by new performance criteria in the form of standards (Holland, 2014).

Because the collapse of the Roman Empire—a major phase transition—set the stage for subsequent events, my investigation begins with the aftermath of Rome's demise. Linked as the Empire was by multiple, intersecting connections, I view Rome's problems and predicaments, its stresses and strains, as being intertwined by links that, with each mishap, set off chain reactions, reverberating throughout the system, and bringing it to a state of collapse in the West (Buchanan, 2002, 226–227; Homer-Dixon, 2006; Riddle, 2016; Wickham, 2017, 23).

5.3   Networks, Standards, and Ties

How did this sea change unfold? To understand this process of reorganization, we need to think in terms of networks and the standard protocols that both configure and tie them together (Buchanan, 2002; Monge and Contractor, 2003; Easley and Kleinberg, 2010; Borgatti et al., 2018). As Grewal (2008) describes:

> A standard defines the particular way in which a group of people is interconnected in a network. It is the shared norm or practice that enables networks to gain access to one another, facilitating their cooperation. A standard must be shared among members of the network to a sufficient degree that they can achieve forms of reciprocity, exchange, or collective effort. (32)

Conceived as such, networks provide a platform for interactions and governing structures within a territorial location or virtual world. By establishing horizontal ties with neighboring networks, standards allow networks to merge and expand outward (Kontopoulos, 1993; Johnson, 2010). At the same time, extending ties vertically via additional standards increases network density and hence the flow of information throughout the network. Increasing ties give rise to new and more diverse inputs, which are required for network sustenance and adaptation (Holland, 2014). As importantly, when networks are linked together in multiplex structures, they cooperate and compete for resources, changing the overall requirements for fitness. In fact, as described below, it was the reconfiguration of network ties that generated phase transitions and their aftermaths, transforming Europe in the long run from a constellation of fluid entities into coherent states (Blanning, 2003 5; Munck, 2005; Padgett, 2016).

5.4 Legitimacy and Role Relationships

Network standards, by themselves, are not enough to bind a social network together. These ties must be reinforced and perceived to be legitimate (Weber, 1964; Bendix, 1996). While some theorists, such as Rawls, define legitimacy in a specific, normative sense—as, for example, in terms of justice (Rawls, 1999), I conceive of legitimacy as an intersubjective, emergent phenomenon, determined through interaction, much like the value of money as described in Chapter 7. In this sense, legitimacy is a derivative of the standard values and norms that constitute culture and define the fitness landscape (Bukovansky, 2002; Brown, 2011, 1–25). When behavior deviates significantly from standard values and norms, legitimacy crises can trigger phase transitions (Bendix, 1996).

To capture this interactive feature of legitimacy, I view legitimacy in the context of role theory and symbolic interactionism (Katz and Kahn, 1978; Biddle, 1986; Collins, 2004). Formulated in these terms, a state's legitimacy can be said to be maximized when its role performance matches, or is congruent with, its constituents' role expectations. Key among these expectations during the Middle Ages were royal bloodlines, land ownership, access to fighting forces, prowess in warfare, displays of wealth, and comity with the papacy. Where gaps in expectations occurred, authorities employed props such as standardized signs, symbols, and rhetoric to bring these expectations into line. Such props were especially important in enhancing the legitimacy of the nobility, sovereigns, and popes during the Middle Ages.

5.5 Shattered Ties in the Early Middle Ages

Between 300 and 700 AD, Europe experienced a "great migration." Barbarians from the East, fleeing the Huns and searching for a better existence, began roaming the European continent. Initially linked together by strong clan ties, a number of their leaders severed these connections during their long trek west and set out to establish dynasties of their own (Heather, 2010; James, 2014).[7]

The barbarian Odovacar was one such leader. In 476, he successfully executed a coup and set off a phase transition by ousting the young Emperor Romulus Augustulus and pronouncing himself King.[8] It was thus that barbarian strongmen acquired the fragments of the empire (Cantor, 2015; Wickham, 2017, 24). Chaos ensued, as the Barbarians established their claims in the course of numerous battles, alliances, marriages, murders, etc. (Halsall, 2008; Bisson, 2009; Cantor, 2015; Blockmans and Hoppenbrouwers, 2017).

Governing the heterodox populations of Europe required legitimation and a stable social structure. Not having an institutional tradition of their own, Barbarians borrowed Roman institutions, incorporating Roman titles, administrative structures, and styles of coinage into their domains (Halsall, 2008, 498–498; Brown, 2011, 34; Riddle, 2016, 93; Wickham, 2017, 26). They sought, moreover, to gain Constantinople's approbation for their plunder and land claims by linking them to the needs of the Eastern Empire. As important, they tied their kingdoms' fates to Christianity and the Catholic Church, gaining their support by defending the papacy and proselytizing Catholicism (Brown, 2011, 15).

Despite these linkages, Barbarians never saw themselves as Romans (Wickham, 2017, 27). Rather, they employed their connections to acquire the same wealth, land, and status as aristocratic Romans, even as they maintained their own distinct cultures (Duby and Clark, 1978; Riddle, 2016, 93; Blockmans and Hoppenbrouwers, 2017, 67; Wickham, 2017, 27). Thus, the new rulers were not inclined to rebuild an empire. For, unlike Rome, which was governed by a bureaucratic state and universal laws, Barbarians ruled based on interpersonal ties (Duby and Clark, 1978, 23–24; Halsall, 2008; Bloch, 2014, 163; Riddle, 2016; Wickham, 2017). Law and order centered on the overriding importance of plundering and vanquishing enemies, as well as honor and loyalty (Duby and Clark, 1978; Bisson, 2009). Justice was administered through the practice of the vendetta, or blood feud (Brown, 2011). Absent overriding authorities to execute justice, individuals pledged themselves to powerful strong men who shielded them from the maelstrom (Bisson, 2009; Cantor, 2015; Riddle, 2016; Wickham, 2017). Thus, post-roman society was fractious and highly uncertain. Above all, it lacked common standards. Consolidation of territory depended on strong men and their warriors, who defeated their competitors and claimed the right to kingship (Crouch, 2005; Halsall, 2008; Riddle, 2016). As Duby (1978) noted:

> In this human void, space was plentiful. What constituted the real basis of wealth was not ownership of land but power over men, however wretched their condition … (13)

5.6 Ruptured Ties in the Merovingian Dynasty

The Merovingian dynasty manifested this state of affairs. At 15 years old, Clovis I (466–511) established the Frankish Kingdom in 481[9] (Riddle, 2016, 104; Wickham, 2020). Although the kingdom encompassed the land from the Pyrenees to Central Europe and ruled for two centuries, it was a contested battleground, characterized by rivalries and perennial warfare among competing Merovingians (Wickham, 2020, 114). Because rulers were obliged by Salian law to divvy up their possessions equally among male offspring, recurrent internecine warfare undermined royal authority. To stay in power, Merovingian kings recruited a vast entourage of warriors, rewarding them with land and titles (Riddle, 2016, 107–108). Most important were the maiores, who exerted considerable influence over Merovingian kings, as they exulted in their luxuries and took little interest in their holdings, delegating power to prominent local, aristocratic landowners who rebelled against attempts to centralize power (Riddle, 2016, 110; Wickham, 2020, 118–119).

Not surprisingly, the Merovingians did little to establish unity. Their monetary policies were totally underdeveloped, so economic ties were sparse (Duby and Clark, 1978, 61–62).[10] In fact, by the end of the 6th century, copper and silver coinage were discontinued, and gold coins had deteriorated so much that they no longer served any commercial purpose (Spufford, 1988 14). Likewise, transportation and communication links collapsed due to the Merovingians' lack of authority and financial wherewithal to rebuild the infrastructure (Blanning, 2003, 18–49). Thus, at the end of the 7th century, only one main route connected Europe to the Middle East, limiting long-distant trade to internal waterways where traders from diverse locations could converge[11] (McCormick, 2020, 20–21). Interregional ties were similarly scarce as struggling local economies aimed to meet their own needs with little left for commerce and trade (Blanning, 2003, 18–49; Wickham, 2017, 37). Also inhibiting trade were local landowners and magnates imposing tolls and customs monopolies, as well as the lack of market information (Munck, 2005, 148–150).

It was under these conditions that Pippin III (the Short), in 749, assumed his father's— Charles Martel's (the hammer's)—position as mayor of the Merovingian dynasty. With the Merovingians in abstentia Pippin—given the Merovingian's lackadaisical attention to their royal duties—called on Pope Zachary to transfer the kingdom to him. The Pope agreed. Hence, another phase transition: by casting asunder traditional ties, Pippin's son Charles subsequently became Charlemagne the Great (Bennett and Bardsley, 2020, 118–119).

5.7 Horizontal Ties in Charlemagne's Regime

Only with the rise of Charlemagne did a semblance of an imperial order reappear (Cantor, 2015; Wickham, 2017, 61–75). However, linked primarily through horizontal ties at the highest level, the Carolingian Empire thrived only so long as Charlemagne's personal connections and authority remained intact. Resolving disputes among his grandsons led not only to the permanent dissolution of the Empire but also, over the long run, to the establishment of independent, sovereign kingdoms.

In contrast to the "do nothing" Merovingians, Charlemagne's Empire (750–814) was impressive indeed. An ardent, ferocious warrior, Charlemagne extended his territory to an area greater than the Roman Empire. Moreover, linking his rule to the Catholic Church, Charlemagne buttressed his own legitimacy while helping to establish Christianity throughout his realm. Together with the Church, he increased literacy and erudition and engaged with intellectuals and the clergy to promote Church reforms (Blockmans and Hoppenbrouwers, 2017).

In contrast to his predecessor, Charlemagne sought to consolidate his power and link his territories by establishing a unified monetary policy. Merovingian coinage had mirrored local political authority because local magnates were authorized to mint coins. Coins were crudely made, and their designs differed according to the tastes of the moneyers and the smiths' skill sets (Spufford, 1988, 43). Taking control of the mint, Charlemagne standardized his coinage in 793, which led to its extensive usage.

While Charlemagne's conquests and monetary reforms unified his empire at the highest level, through the promotion of horizontal ties, his efforts to extend power downward via local ties proved less effective (Bradbury, 2007, 12; Cantor, 2015, 129–130). To reinforce his claims to power, Charlemagne had to expand his territory by conquest so as—in the absence of freely circulating coinage—to reward his vassals with land (Spufford, 1989, 16; Blockmans and Hoppenbrouwers, 2017). Thus, the growth of the empire precluded direct control from above. Although Charlemagne empowered underlings to supervise and monitor local affairs, these appointees, which were often selected on an interpersonal basis, lacked administrative skills and experience[12] (Cantor, 2015, 17; Bennett and Bardsley, 2020, 131–132; Blockmans and Hoppenbrouwers, 2017). And, increasingly, they were less likely to give up control.

The 9th century was especially stressful for the Carolingian dynasty. Not only was it threatened by the invasions of the Scandinavians, Magyars, and Moslems, but also by the conflicts among Charlemagne's heirs. In keeping with Salian law, Charlemagne bequeathed his empire to his son Louis, upon whose death it was passed on to his three grandsons—Louis the Pious, Lothar, and Louis the German. As might be expected, given the contentious political climate at the time, his grandsons began to encroach upon each other's domains. On inheriting their portions, Charlemagne's great-grandchildren followed suit. Hence, fighting was pervasive, and the empire continued to disintegrate (Blockmans and Hoppenbrouwers, 2017).

5.8 Consolidation with Re-Emerging Ties

Porous boundaries allow networks to merge, and thereby gain greater skills and knowledge, allowing agents to move higher up the fitness landscape (Holland, 2014). Such was the case throughout the Middle Ages, when European populations migrated across the continent, and, after fiercely and ferociously conquering new lands and cultures, became subsumed and incorporated within them. The result was greater competence and more complex and richer heritages. The first stage of this consolidation, which constituted a major phase transition, coincided with the arrival of the Vikings, who uprooted the status quo.

Towards the end of the Carolingian period, the Vikings, having colonized Iceland, Greenland, and Ireland, arrived in superbly crafted boats to attack England's northeast coastlines. By all accounts, Vikings were brutal and murderous warriors, pillaging and plundering coastal towns and slaughtering their inhabitants (Sawyer, 1982; Winroth, 2004). Facing little resistance and reaping considerable bounty, the Vikings returned to raid again at subsequent intervals. In 865, they mounted an attack on England with a large army, seeking to establish permanent settlements. Defeated ten years later by King Alfred of Wessex, the Danes reached an accommodation establishing a border between English and Viking territories, known, thereafter, as the Danelaw. But then, Sveyn Forkbeard, son of the Danish King Harold Bluetooth, conquered most of England in 1013. Ruling for only five weeks, Sveyn was succeeded by his son, Cnut the Great (1016–1035), who pronounced himself King of England as well as King of the Danes and the Norwegians. (Winroth, 2004; Bloch, 2014, 38–39; Blockmans and Hoppenbrouwers, 2017, 181–185). King Cnut ruled England for two decades, followed by his two sons. Upon their deaths, the crown fell once again to the English under Edward the Confessor (Winroth, 2004). Despite their previous hostilities, the Saxons and the Danes were eventually assimilated together with the English, as evidenced by the Danish conversion to Christianity and the addition of many Danish linguistic features into the English language.

The Vikings also pillaged France, beginning in the 9th century. King Odo (988–998), the first non-Carolingian king, successfully repelled the Vikings a number of times. However, given repeated attacks, Odo called for aid not only from his vassals but also from Vikings already settled in Francia. As raids continued into the 10th century, Odo's successor, Charles the Simple, turned to the Viking Rollo, bestowing upon him Normandy, on the condition that he defend the territory; do homage to Charles; and convert to Christianity (Bradbury, 2007, 33–34; Bloch, 2014, 42–43). Putting down roots, the Vikings in France thereupon renounced aggression in favor of trading, farming, and establishing families of their own.

All in all, Viking assimilation within Europe was far reaching. Having reached their goals in the West, many returned to Scandinavia, bringing with them much of the practical knowledge they had gained abroad (Bloch, 2014, 51). It was thus that the Vikings experience in Western Europe paved the way for Scandinavian incorporation into the European orbit. As complexity theory might define it, here was an example of adaptation generating increased fitness, on both sides, due to sharing across the adjacent possible. As Winroth (2004) attests:

> The Viking Age not only brought Scandinavia into concrete and direct contact with the rest of Europe [;] it also took the region into the European mainstream. When Charlemagne's courtiers jested in the 780's about converting "impious" "brute" and "impenetrable" Danish kings to Christianity, Scandinavia was still far outside European civilization and culture, and the Franks could only joke about the region…. After the end of the Viking Age, in the twelfth and thirteenth centuries, Scandinavia was part of Europe, no longer an alien region beyond civilization but a region that was organized and structured along the same lines as the rest of the continent. (169)

5.9   Free-For-All, Due to Lack of Standardized Ties

Violence was endemic in the years following the collapse of the Carolingian Empire. For the Vikings were not the only ones marauding throughout Europe (Kaeuper, 2002). During the fractious 11th century, a new contingent of warriors emerged. These were knights—"professional," warriors who provided armed services for Kings and Lords, and who, in their ancillary time, fought with one another as well as the right to carry out private wars (Kaeuper, 2002). Relishing their identity as warriors, they loved to fight for the sake of fighting, and so they put themselves up for hire. As Kaeuper (2002) notes:

> The fighting, let us remember, was not merely defensive. Not simply carried out at the royal behest in defense of recognized national borders, not only on crusade, not really (despite their self-deceptions) in the defense of widows, orphans, and the weak … They fought each other enthusiastically as any common foe; perhaps even more often they brought violence to villagers, clerics, townspeople, and merchants. (11)

Given the weakness of higher authorities and the lack of common standards, public order was dreadfully lacking and verging on chaos. In the absence of a working authority, local magnates competed to seize land, build castles, and acquire ruling titles. Characterizing this devolution of power, Bennett and Bardsley (2020) explain:

> As the monarchy became a far-off irrelevant institution, dukes and counts changed from men empowered by royal office into territorial princes only tied loosely to their king. Powers they had held by grant of the king—custody of royal lands, tribunals, tax; revenues, and military conscription—ripened into hereditary possession they held by virtues of their father and mothers. Lands and powers they had once administered for the king they now administered for themselves. The heartland of Charlemagne's empire was transformed into a mosaic of hundreds of largely independent duchies and counties. (148)

To defend their claims, these burgeoning lords sought knights to become their vassals and fight on their behalf. These knights not only defended their lords; restless in times of peace, they also fought each other, and, seeking adventure, they travelled long distances to compete in perilous jousting tournaments, where they could hone their skills and prove their mettle. Prowess was, above all, a measure of knightly status (Tuckman, 2014).

Notwithstanding their penchant to engage in combat, knights saw themselves as colleagues of sorts—an elite group, distinct from others in society. Standards, in the form of a chivalric code, reinforced their collegiality and guided their behavior.[13] The code of chivalry, developed in the 12th century, called upon knights to exhibit prowess, loyalty, steadfastness, forbearance, and largesse, among other qualities (Crouch, 2005). These standards were reinforced and elaborated upon in the widely shared Arthurian legends and the chansons de geste, which glorified knightly adventures.

Providing high-status, knighthood initially drew men from all realms of life. For younger men, it was a means of social mobility, opening up the possibility of new connections and opportunities. Those whose performance eclipsed others might be granted a fiefdom and a noble title (Blockmans and Hoppenbrouwers, 2017, 258–266). Even the prelates of the Church, who were repelled by excessive violence, recognized the need for knights to whom they looked to protect their monasteries and go on crusade. To tame knights' violent proclivities, Church prelates sought to infuse piety into the chivalric code, as well as to clearly distinguish between religious wars and those waged for plunder and conquest (Kaeuper, 2002; Blockmans and Hoppenbrouwers, 2017, 281). As s knighthood flourished, those in the top ranks of society aimed to restrict this status to themselves. Hence, over time, chivalry became associated with the nobility. As Goodman (1992) has described:

> Chivalry [is] a code of conduct originally developed for a tightly defined professional group, but later appealing to figures beyond the boundaries of the original chivalric world; the prince, the landed gentry the professional soldier, sometimes even the merchant. Because of its prestige as a system of values, the European chivalric ideal in fact outlast[ed] the social class that created it. (5)

5.10   Dense Ties, and the Retreat into Feudalism

The shift of power to landed aristocrats brought about another phase transition, as a new form of social and economic organization emerged—namely, feudalism (Duby and Clark, 1978, 161; Lopez, 1976; Bloch, 2014). The system of feudalism, prevailing from the middle of the 9th to the first decades of the 13th century, was a land-based system of social and economic life that took the form of a hierarchical, self-sufficient, organic whole, centered around the manor (Lopez, 1976, 49, Blockmans and Hoppenbrouwers, 2017, 152–153). The manor provided sustenance and security to its subjects, while its subjects provided services in-kind to the lord. Status, defined by rigid standards, prescribed each member's distinct but complementary role, making the manor relatively self-sufficient overall. Wealth was accrued by exploiting peasants and serfs in accordance with a wide range of complex norms and legal obligations (Duby and Clark, 1988; Bloch, 2014, 351–352; Wickham, 2017). Hence, wealth was in-kind, and coinage mostly irrelevant. Although the feudal system resolved the problems of coordinating supply and demand in the post-Carolingian era, it had drawbacks due to its inflexible and unproductive manner. Because the level of manpower was relatively fixed, as was the cost of sustenance, the lord could not adjust his priorities as economic conditions changed. And change they did, leading to a subsequent phase transition.

Money, according to Spufford, was determinant in the 13th-century phase transition (Spufford, 1988, 1). Above all, money upended the aristocracy's relationship to their peasants and serfs, upon whom they depended for subsistence. When monetary rents replaced in-kind rents, peasants gained greater independence and opportunity. However, once on their own, peasants had little recourse in times of trouble. Not surprisingly, they bore the brunt of the late 13th-century famines, wars, and plagues. Moreover, although population decline led to higher wages for some, many peasants lost their lands and livelihoods. The sundering of traditional relationships between peasants and landowners, together with peasants' increased awareness of the gaps in prosperity, gave rise to peasant revolts across Europe, as well as their migration to emerging cities (Bennett and Bardsley, 2020, 175).

As telling, feudalism's maximization of local ties at the expense of long-distance ties foreclosed new, lucrative trading opportunities, which emerged with the proliferation of money. To overcome these barriers, landed aristocrats established new, extended ties by sponsoring European-wide fairs, where producers, merchants, and bankers could congregate from long distances (Braudel, 1992). Providing platforms of interlinked ties between merchants, buyers, and moneylenders alike, interregional trade fairs reduced transaction costs and generated economies of agglomeration, such that everyone gained. Landowners benefited from considerable revenues drawn from taxes and tolls. Merchants obtained greater sales and market information, while bankers profited from new financial standards—such as bills of exchange—which provided a source of credit that, circulating together with the sale of goods, bridged the long trading cycle. (Braudel, 1992; Nicholas, 2014, 30–31). Given these innovations, trade, and the diverse ties it generated, were extended widely in the 15th century to the Far East and the New World. These developments, together with the organizational restructuring associated with the rise of commercial cities, rendered asunder feudal ties in a phase transition.

5.11 Overlapping Ties and Complexity Catastrophes

With the loosening of feudal ties, nobles came into their own. Given greater resources, they sought to enhance their dynasties by both increasing the size of their territories and deepening their authority within them (Watts, 2009, 106–111). To legitimate their actions, nobles expanded their domains in areas where they had some historical continuity. However, given the fluid geography of the Middle Ages, many areas were contested by others who had competing claims (Wale and Denley, 2013, 36–37). Thus, despite nobles' ambitions, realizing their aims risked the onset of war. Under the circumstances, new standards designed to link political actors together and lessen the risk of conflict emerged. However, these ties not only grew in number; they also increasingly overlapped, creating a complexity catastrophe (Kauffman, 1993, 1995).

Complexity catastrophes are a major source of phase transitions (Kauffman, 1995). They occur when ties not only overlap but are also tightly coupled. Hence there is little slack, or room for maneuver, so opportunities for conflict resolution are extremely limited. In the late Middle Ages, complexity crises occurred when cross-cutting interconnections between kings, lords, and prelates engendered conflicting goals and competitive relationships. Ironically, these overlapping ties stemmed from social practices, such as homage and marriage alliances that, although intended to ensure peace and stability, inadvertently contributed to the buildup and execution of a major phase transition—the Hundred Years War (1337–1453).

In contrast to manorialism, which defined the relationships between peasants and their overlords, homage was a social standard that governed political linkages between nobles, lords, and vassals. The practice of homage can be traced back to the Carolingian era, when lords, battling one another in highly overwrought environments, needed vassals to fight on their behalf. Although homage relationships between lords and vassals were hierarchical—requiring vassals' submission to their lords—they were based on, and reinforced by, strong personal bonds, which both parties freely entered into (Neillands, 1990; Reynolds, 2001[14]; Watts, 2009, 89; Blockmans and Hoppenbrouwers, 2017, 144). In pledging their fealty, vassals promised to fight for their lord against his enemies; to provide him with good council; and to otherwise do his bidding.[15] In exchange, lords swore to protect their vassals and to provide for their well-being, a commitment that might entail the granting to him of a fief, as well as the associated governing duties, as was the case of France.[16]

This pattern was reiterated on downward from one vassal to the next. For, once a vassal was enfeoffed, he might engage vassals of his own (Watts, 2009, 89; Waley and Denley, 2013, 34–35). Given these complex configurations and blurred political boundaries, a vassal might be a powerful king, a duke, or a lord who surpasses the power and wealth of his own overlord[17] (Neillands, 1990, 29; Watts, 2009, 61–62, 83–84). Such was the case, for example, when Henry

II married Eleanor of Aquitaine, providing the King of England wealth and power exceeding that of the King of France, his feudal overlord (Neillands, 1990, 18). Alternatively, a vassal might owe fealty to two or more lords, who are at odds with one another, raising questions not only about which lord should take precedence, but also about how to avoid committing treason under the circumstances. Equally paradoxical was the situation when the King of England found himself at war with France, even as he had pledged homage to its king. It was, in fact, just such situations involving competing loyalties and claims of homage that, instead of preserving peace and ensuring stability, generated complexity catastrophes leading to warfare (Neillands, 1990).

Ties among families and alliances exacerbated these problems. As Crouch points out, the "family" was the basic component of medieval society (Crouch, 2005). In many cases, family relationships were governed by Salic Law, which was designed to perpetuate a family's lineage as well as the wealth, status, and other trappings related to it. To preserve a noble's family lineage, the law provided that the succession of kings was restricted to the oldest member of the male line of a family.[18] If no males assumed this role, the lineage would die out, as it did when the Valois replaced the Capetians in France. Although Salic Law had a long history dating back to the Merovingians, it often proved a source of conflict, not only among competing monarchs but also among the rivalrous offspring of kings. One need only consider the hostilities and clashes over land and power that occurred among Charlemagne's grandchildren upon the death of their father, Louis the Pious. Likewise, late in Henry II of England's realm, his three oldest sons, along with their mother, Eleanor, shifted their allegiance to his father's enemy, only to be subsequently reconciled with the English King (Jones, 2014). Thus, although family ties and their alliances were intended to foster peace and security, they were just as likely to instill conflict.

5.12   Cities and the Agglomeration and Diversification of Ties

Cities not only contributed to the decline of feudalism by channeling populations and economic activities in their direction; even more importantly, they generated new lifestyles based on a wide range of new standards and resources (Riddle, 2016, 67–68; Nicholas, 2014). Serving as platforms, cities functioned as magnets, attracting an ever-increasing number of resources, including human talent, financial investments, diverse specialized products, inhabitants, and other resources. As Nicholas has described, 11th and 12th century towns became "an economic agglomeration that not only consumed but also produced exportable industrial goods and provided important services of reconsignment of basic necessities as well as luxuries" (Nicholas, 2014, 19).

Lords and bishops were the first city dwellers, along with their entourages. Soon, serfs and peasants arrived to provide food and other necessities in exchange for the coinage needed to pay their rents. Next, many aristocrats left their country estates in favor of city life (Riddle, 2016; Nicholas, 2014, 21). No longer dependent on in-kind rents, aristocrats engaged in the city's courtly life, spending lavishly on high-status goods, such as silks, spices, and precious jewels (Spufford, 1988; Wickham, 2020). To meet the demand for high-value goods, long-distance traders—such as Jews, Syrians, and Venetians—ascended on local towns. Describing the evolution of monetary interactions, Hunt and Murray (1999) note:

> This demand for luxuries thus both began and completed the circle, for if peasants supplied the surplus, it was the merchants who converted it into money, traded the money for the

desired luxury goods, sold the item to the lord, and profited from the role of entrepreneur. The first great ages of European history were nothing more than a nearly infinite number of variations on this simple theme. (28)

As towns flourished, cities renegotiated their relationship with elites, establishing charters that freed them from feudal obligations. Kings were more inclined than lords to grant such charters because thriving cities counterbalanced the growing power of independent lords. Absent feudal rule, town leaders established communes, whose leadership was drawn primarily from town merchants. These merchants not only fostered the growth of markets, allowing trade to take place between diverse products and specialized goods; they also forged new links to trade routes and trading partners (Spufford, 1988, 1).

Key to this growth was their outreach to overseas settlements, allowing commercial agents to gain critical market information while dispensing with foreign middlemen (Banaji, 2020, 14). In the process, cities emerged as the driving force of the European economy (Braudel, 1992). What fueled their momentum was not only the acquisitive modus operandi of their merchants but also the extent of their network connections, which—made up of both dense, local links and long-distance ties—constituted a "small world"[19] with all its benefits. Projecting themselves out from their city-centered economies, these cities extended their grasps far afield by colonizing parts of Asia and North and South America (Brown, 2011).

5.13 Church vs. State, Co-opetition at the Top

The Church was a major centerpiece of the Middle Ages. It was a key source of the standards governing the era, and a critical power broker determining how these standards were to be executed. It is not surprisingly, therefore, that conflicts throughout the Middle Ages gave rise to phase transitions that spilled over from the religious realm to all other aspects of medieval life. For, as we learn from complexity theory, when two networks consistently interact, they not only cooperate; they also compete. In so doing, they change the fitness landscape and the overall fitness requirements across all interconnected networks (Holland, 2014).

Throughout the Middle Ages, the institutions of the Church and States were not only interlinked, they were co-dependent. The monarchy relied on the Church to bolster its legitimacy, while the Church relied on Kings and Lords to enhance its political/military powers. Thus, the Church turned to Kings and Lords to fight in the crusades, whereas the kings sought spiritual confirmation as well as symbolic legitimacy from their investitures at the hands of the Pope (Southern, 1990). Nevertheless, tensions were rife between them (Maland, 1982, 126). At the heart of their disputes was the claim that, each's powers were supreme and prescribed by God within their own, separate realms. However, over time, this postulation proved problematic.[20] As the social order became more complex, the boundaries between the sovereign realm and the spiritual realm began to overlap, and major conflicts about appropriate rights and duties of each soon followed (Southern, 1990, 163–168).

As complexity theory might anticipate, the interdependencies between the Church and State mirrored the overall fitness landscape. So long as both pillars of power had clear-cut domains and relatively equal—albeit distinct—resources to leverage against one another, they typically refrained from employing them to the fullest extent. Needing each other to balance the political forces at hand, compromise was their only option (Southern, 1990). However, when they were unable to compromise, a phase transition ensued.

5.13.1   The Investiture Crisis and the Treaty of Worms

Network boundaries between Church and State were ill-defined in the early 11th century, when the imperial King of Saxony reached out to bishops and abbots to perform secular roles typically associated with counts. Given their superior education and administrative expertise, incorporating high-level clergy members into royal affairs had many benefits. However, for such a system to function successfully, the king had to have the right to appoint or, at the very least, approve members of the clergy (Blockmans and Hoppenbrouwers, 2017, 200). When disagreements about power sharing failed, conflict followed.

One need only consider the case of King Henry IV. To strengthen his position as Emperor and King of Saxony, King Henry IV, in 1075, appointed the archbishop of Milan and two bishops, who stemmed from towns that were under the Pope's jurisdiction. Pope Gregory VII (1073–1085), an ardent defender of Church rights, was furious. In retaliation, he excommunicated Henry. But Henry, having fragile relations with his own nobles, could not afford such effrontery. Seeking reconciliation with the Pope, he rushed to Canossa to meet him. There, outside the Pope's door, Henry waited in the snow, barefoot and dressed in sackcloth. When the Pope finally received Henry, he was forced—given the separation between Church and State—to forgive him. Hence, this conflict ended in a draw, leaving Henry's opponents marginalized as a result (Blockmans and Hoppenbrouwers, 2017, 159–71). In contrast, in England, as described below, conflicts between the Church and State had far greater impacts, taking the form of a phase transition.

5.13.2   The Canterbury Crisis

In 1162, England's King Henry II sought to reinforce his position vis-a-vis the Church. In particular, he aimed to ensure that, as king, he would have the ultimate say in criminal cases ruled on by the clergy. To this end, Henry looked for support from his former loyal associate and newly appointed Archbishop of Canterbury, Thomas Becket.[21] When Becket demurred, Henry was outraged. Retaliating, he demanded Becket publicly pledge support for the Constitutions of Clarendon—laws that entirely suited the king. Under such pressure, Becket agreed but thereafter retracted his support. Fearing retaliation, Becket fled to France, where he was exiled for six years. Returning to England, Becket resumed his hostile behavior, reigniting the King's fury. Hearing Henry remonstrate about Becket, four of Henry's knights "took the hint" and murdered Becket in the Canterbury Cathedral, an act that made Becket appear to be a martyr and a saint. To atone for his role in this affair, Henry underwent penance in the cathedral where he was subject to whippings before monks and knights and forced to retract the sections of the Clarendon Constitutions that were detrimental to the Church (Blockmans and Hoppenbrouwers, 2017, 222–223). Although the Church won this battle, it lost the extended war by crossing the line from religion to political affairs. No longer operating in separate spheres, church/state relationships evolved into a zero-sum game, so that when the power of the Church subsequently declined, that of the kings and secular magnates only increased (Blockmans and Hoppenbrouwers, 2017, 223).

5.13.3   A Fractured Church and Its Subsequent Decline

Pope Clement's relocation to Avignon in 1309 created a major fissure between the Roman and French factions within the Catholic Church. As a result, the boundaries between Church and State became increasingly blurred, as European kings and despots took sides in the papal rivalries. Exacerbating the discord, Clement was succeeded by six French popes, together with their French entourages, all of whom had significant ties to the French state. In addition to the clergy's reputation for favoring French political interests, the Avignon contingent was notorious for its profligate ways. As Tuckman (2014) describes:

> Avignon became a virtual temporal state of sumptuous pomp, of great cultural attraction, and unlimited simony—that is the selling of offices. Diminished by its removal from the Holy See of Rome and by being generally regarded as a tool of France, the papacy sought to make up prestige and power in temporal terms. (49–50)

Although the French Pope Gregory XI returned to Rome in 1377, the situation continued to deteriorate. Then, in 1387, after the French cardinals rejected the election of Urban VI to the papacy, two competing popes were chosen, one residing in Avignon, the other in Rome. To make matters worse, failed reconciliation between the factions led to the selection of a third claimant, the antipope Alexander V. Then, in 1417, Martin V, an Italian who had supported the antipope, was elected at the Council of Constance, ending the Church schism, which had lasted from 1309 to 1417.

This schism not only affected power relationships within Catholicism; it also impacted the course of European Christianity. Widely followed throughout Christendom, the rivalry among the Catholic prelates and the notoriety surrounding their debauched and corrupt behavior undermined the legitimacy of the Church and reinforced movements calling for reform. It was within this context that, in yet another phase transition, protestant movements emerged, establishing new standards for worship and eliminating the Church's intermediary role between individuals and their God.

5.14 The End of the Middle Ages

The denouement of the Middle Ages can be conceived as the finale of a supervening phase transition. When viewed as a whole, the Middle Ages appears as a series of phase transitions that over time engendered innovative relationships governed by new standards of interaction, opening up what Swiss cultural historian Jacob Burckhardt characterized as "a window into the modern era" and what complexity theorists might view as a major shift to the "adjacent possible" (Burckhardt, 1995). Just as the Middle Ages emerged from, and built upon, the remnants of the Roman Empire, so too the Renaissance drew upon the building blocks cast down from the Middle Ages. Critical to the deconstruction of the Middle Ages were the series of phase transitions that emerged when actors, locked into their dense interrelationships, failed to adapt to major changes in their environments by generating new standards and organizational forms to address them. This string of developments set the stage for the modern era.

Notes

1 Fernand Braudel, a renowned French historian, was a leading proponents of the Annales School of History. Founded by Lucian Febvre and Marc Bloch, the Annales school encouraged historians to be interdisciplinary, focusing on the everyday sociological patterns that led, over time, to major societal changes. See for an account, Burke, 1990.

2 Braudel describes events as the ephemera of history; "they pass across its stage like fireflies, hardly glimpsed before they settle back into darkness and as often as not into oblivion" (Braudel, 1992, 2, 901, as cited in Lee, 2012, 24).

3 The fitness landscape, a term derived from evolutionary biology, is a metaphor that refers to the criteria of fitness in any giving situation. It is depicted as a series of mountain peaks, in which species seek out the higher peaks where fitness is the greatest. All too often, however, searchers get stuck on lower peaks where fitness is suboptimal. The best strategy for achieving high fitness levels is for the species to split up and search separately, on the assumption that some will reach the peaks, reproduce, and preserve the species. For a discussion, see Stewart Kauffman, 1995, 40–41.

4 Critical points are the points between two physical states which are poised to flip given the right circumstances.

5 As Chase-Dunn and Hall have demonstrated in their research on world systems, major, abrupt societal ups and downs have repeatedly occurred throughout human history (Chase-Dunn and Hall, 2019)

6 If ties are too dense and entangled, they will give rise to a complexity catastrophe (see Kauffman, 1993). This is an argument that Herbert Spencer made in the late 1890s when he opined that "populations may not be able to respond to selection pressures because of cultural or institutional rigidities, thereby causing the dissolution of the population and disintegration of its sociocultural formations" (Turner and Machalek, 2019, 42).

7 As Cantor notes: "During this period, the bonds of kinship were weakened, a process that shows itself in the prevalence of strife between relatives. The binding forces formally wielded by kinship was increasingly transferred to the relationship between "lord" and "man," between whom no bond of blood relationship was necessary, only the bond of loyalty "(Cantor, 2015, 68).

8 It is this date that is associated with the collapse of Rome, although the fall of the western empire was less a singular event, and more an unfolding process (Riddle, 2008, 48–50).

9 Clovis became emperor at the death of Theodoric, who—at the urging of the Byzantines—had murdered Odovacar and replaced him as emperor. According to Cantor, Theodoric was second only to Charlemagne in terms of greatness. However, given his prowess, Theodoric aroused the Byzantines' anxiety, especially as he aspired to building an all-gothic empire in the west. The Byzantines were, thus, in favor of eliminating Theodoric (Wickham, 2009; Cantor, 2015).

10 Not having a tradition of coinage and seeking to maintain good relations with the Eastern Empire, Merovingian kings, up until the 6th century, issued gold, and later silver, coins in the Roman style, referencing on them ancient or Byzantine emperors (Spufford, 1989, 9). Likewise, minting was originally carried out on a decentralized basis, in mints that had been established by moneyers in Roman times (Kohn, 1999).

11 In the North Sea and the Mediterranean, small trading emporia emerged in the seventh and early 8th centuries. But, even there, the market structure was ad hoc in nature. As well, the lack of any laws governing market behavior and lawlessness on the roads greatly inhibited trade (Munck, 2005, 148–149; McCormick, 2020, 20–21).

12 According to Blockmans and Hoppenbrouwers, "In many respects, the apparatus of state was limited to the court and a few hundred officials who struggled to impose laws in the enormous stretching from the Ebro to the Elbe and including large parts of Italy. In the absence of a developed tax system, it was a hopeless task to impose lasting administrative unity on the jumbled diversity of peoples belonging to different cultures and levels of development" (Blockmans and Hoppenbrouwers, 2017, 143–144).

13 For over a century, scholars were hesitant to do serious analysis of chivalry because it overlapped with the Arthur legends and other romantic themes, so that it was not considered to be a serious pursuit. However, its impact on the social order can hardly be disputed. (Crouch, 2005, 20–25).

14 Susan Reynolds downplays the role that homage played during the Middle Ages based on her research on feudal laws relating to fiefdom. What concerns us here is not so much fiefdom, as defined by law, but the prevailing view during the period that a vassal owed his person to the lord, an agreement that was committed not by law but by face-to-face oral vows. History documents the numerous cases in which fiefdom had a causal effect.

15 The oath of fealty been vassals and their lords took place in an elaborate ritual in which the vassals placed their hands between the hands of the lord and, upon hearing the pledge of fealty, the lord kissed him on the mouth. At this point, the vassal became the lord's "man."

16 As Watts points out, "… lordship was a highly personal form of rule, which does not mean that it was simply dependent on individual character, but rather that it involved the direct interaction of individuals, the management of individual relationships, and attention to individual needs and misdemeanors" (Watts, 2009, 86).

17 According to Watt, "Kings were frequently ineffective and their interventions were often resented and resisted; other power holders could command allegiance too; other communities could often be more

meaningful that the "regnal" one. A balanced perspective on kingship, which acknowledges its power without assuming a natural preeminence is an essential basis for understanding of the politics of [the] period "(Watts, 2009, 88–89).

18 An exception was made in the Holy Roman Empire, where successors to the Empire were chosen by election. As Waley and Denley describe: "Elective kingship meant, above all, lack of continuity. This was a tremendous disadvantage in administration. Each time a new feudal family assumed the coveted burden of imperial rule, new officials took over and a new archive was set up (Waley and Denley, 2013, 121).

19 Small worlds are constituted by both short and long-distance ties. The long-distance ties provide a short cut across a geographic region, allowing it to be crossed by just six steps. At the same time, short ties foster the levels of trust required for long distance trade (see Buchanan, 2002).

20 As Southern attests: "Indeed it is evident that the idea of [building] a western empire [by means of] extending papal authority was mistaken from beginning to end. It was a mistake primarily because in creating an emperor the pope created not a deputy, but a rival or even a master. The theoretical supremacy implied in the act of creation could never be translated into practical obedience to order given or received. Hence the Pope's practical supremacy came to an end at the moment of coronation (Southern, 1990, 166).

21 Thomas Becket had previously served Henry in the role of Chancellor. Although Becket had assiduously supported Henry's policies in that role, he had somewhat of a tarnished reputation. Once he had been appointed Archbishop, he transferred his support to the church. He was a stout believer that the monarch should not be involved in church affairs. Historians suggest that Becket's transformation resulted from his doubts about his own capabilities, for which he greatly overcompensated (Blockmans and Hoppenbrouwers, 2017, 222–223).

References

Bak, Per (1996) *How Nature Works: The Science of Self-Organized Criticality*, New York, NY: Springer Science & Business Media.

Banaji, Jarius (2020) *A Brief History of Commercial Capitalism*, Chicago, IL: Haymarket Books.

Bendix, Reinhardt (1996) *National Building and Citizenship: Studies of Our Changing Social Order*, New York, NY: Routledge.

Bennett, Judith, and Sandy Bardsley (2020) *Medieval Europe: A Short History*, New York, NY: Oxford University Press.

Blanning, T.C.W. (2003) *The Culture of Power and the Power of Culture: Old Regime Europe, 1660-1789*, New York, NY: Oxford University Press.

Biddle, Bruce J. (1986) "Recent Developments in Role Theory," *Annual Review of Sociology* 12, 67–92.

Biddle, Bruce J. (2013) *Role Theory: Expectations, Identities, and Behaviors*, New York, NY: Academic Press.

Bisson, Thomas N (2009) *The Crisis of the 12th Century: Power, Lordship, and the Origins of European Government*, Princeton, NJ: Princeton University Press.

Bloch, Marc (2014) *Feudal Society*, New York, NY: Routledge.

Blockmans and Hoppenbrouwers (2017) *Introduction to Medieval Europe*, 300–1500, 4th edition, New York, NY: Routledge.

Borgatti, Stephen, et al. (2018) *Analyzing Social Network*. New York, NY: Sage Publications.

Bradbury, Jim (2007) *The Capetians, Kings of France*, 987–1238, London: Continuum Books.

Braudel, Fernand (1992) *The Wheels of Commerce*, Berkeley, CA: University of California Press.

Brown, Warren (2011) *Violence in the Middle Ages*, New York, NY: Routledge.

Buchanan, Mark (2002) *NEXUS: Small Worlds and the Groundbreaking Science of Networks*, New York, NY: W.W, Norton and Company.

Buchanan, Mark (2007) *The Social Atom*, New York, NY: Bloomsbury.

Bukovansky, Mlada (2002) *Legitimacy and Power Politics*, Princeton, NJ: Princeton University Press.

Burke, Peter (1990) *The French Historical Revolution: The Annales School, 1929–1989*, Cambridge: Polity Press.

Cantor, Norman F (2015) *Civilization of the Middle Agers: A Completely Revised and Expanded Edition*, New York, NY: Harper Collins Publisher.

Chase-Dunn, Christopher, and Thomas D. Hall (2019) *Rise and Demise: Comparing World Systems*, New York, NY: Routledge.

Collins, Randall (2004) *Interaction Ritual Chains*, Princeton, NJ: Princeton University Press.

Crouch, David (2005) *The Birth of the Nobility: Constructing Aristocracy in England and France, 900–1300*, New York, NY: Routledge, 2014.

Curtis, Benjamin (2013) *The Hapsburgs: History of a Dynasty*, New York, NY: Bloomsbury Academic.

Duby, George, and Howard B. Clark (1978) *Early Growth of the European Economy: Warriors and Peasants from the Seventh to the Twelfth Century*, Ithaca, NY: Cornell University Press.

Easley, David, and Jon Kleinberg (2010) *Networks Crowds and Markets: Reasoning about a Highly Connected World*, Cambridge: Cambridge University Press.

Goodman, Jennifer (1992) "European Chivalry in the 1490s," *Comparative Civilizations Review* 26 (26).

Grewal, David Singh (2008) *Network Power: The Social Dynamics of Globalization*, New Haven, CT: Yale University Press.

Halsall, Guy (2008) *Barbarian Migrations and the Roman West*, 376–569, Cambridge: Cambridge University Press.

Heather, Peter (2010) *The Fall of Rome and the Birth of Europe*, New York, NY: Oxford University Press.

Holland, John (2014) *Signals and Boundaries: Building Block for Complex Adaptive Systems*, Cambridge, MA: MIT Press.

Homer-Dixon, Thomas (2006) *The Upside of Down, Catastrophe, Creativity, and the Renewal of Civilization*, New York, NY: Random House.

Hunt, Edwin S., and James M. Murry, (1999) *A History of Business in Medieval Europe*, 1200–1550, New York, NY: Cambridge University Press.

James, Edward (2014) *European Barbarians AD 200–600*, New York, NY: Routledge.

Johnson, Steven (2010) *Where Do Good Ideas Come From: The Natural History of Innovation*, New York, NY: Bloomberg.

Jones, Dan (2014) *The Plantagenets*, New York, NY: Penguin.

Kauffman, Stewart (1993) *The Origins of Order: Self-Organization and Selection in Evolution*, New York, NY: Oxford University Press.

Kauffman, Stewart (1995) *At Home in the Universe: The Search for Laws of Self Organization and Complexity*, New York, NY: Oxford University Press.

Katz, Daniel, and Robert L. Kahn (1978) *The Social Psychology of Organizations*, New York, NY: John Wiley and Sons.

Kaeuper, Richard W. (2002) *Chivalry and Violence in Medieval Europe*, New York, NY: Oxford University Press.

Kontopoulos, Kryiakos (1993) *The Logic of Social Structure*, New York, NY: Cambridge University Press.

Lee, Richard (2012) *The Longue Durée and World System Analysis*, New York, NY, Suny Press.

Lopez, Robert S. (1976) *The Commercial Revolution of the Middle Ages*, 950–1350, New York, NY, Cambridge University Press.

Maland, David (1982) *Europe in the Sixteenth Century*, New York, NY: Macmillan Press.

McCormick, Michael (2020) "Discovering the Early Medieval Economy," in Davis, Jennifer, R., and Michael McCormick, eds. *The Long Morning of Medieval Europe*, New York, NY: Routledge.

Monge and Contractor (2003) *Theories of Communication Networks*, New York, NY: Oxford University Press.

Munck, Thomas (2005) *Seventeenth Century Europe, State, Conflict and Social Order*, New York, NY: Oxford University Press.

Neillands, Robin (1990) *The Hundred Years War*, New York, NY: Routledge.

Nicholas, David (2014) *The Later Medieval City, 1300–1500*, New York, NY: Routledge.

Pennington, D. H. (2014) *Europe in the 17th Century*, New York, NY: Routledge.

Rawls, John (1999) *A Theory of Justice*, Cambridge, MA: MIT Press.

Reynolds, Susan (2001) *Fiefs and Vassals: The Medieval Evidence Reinterpreted*, New York, NY: Oxford University Press.

Riddle, John M. (2016) *A History of the Middle Ages*, 300–1500, 2nd edition, New York, NY: Rowman and Littlefield.

Sawyer, Keith (2005) *Social Emergence: Societies as Complex Systems*, New York, NY: Oxford University Press.

Sole, Richard (2011) *Phase Transitions*, Princeton, NJ: Princeton University Press.

Southern, R.W. (1990) *Western Society and the Church in the Middle Ages*, New York, NY: Penguin Books.

Spufford, Peter (1988) *Money and Its Use in Medieval Europe*, New York, NY: Cambridge University Press.

Tuckman, Barbara (2014) *A Distant Mirror*, New York, NY: NYL Random House.

Turner, Jonathan H., and Richard H. Machalek (2019) *The New Evolutionary Sociology: Recent and Revitalized Theories and Approaches*, New York, NY: Routledge.

Waley, Daniel and Denley, Peter (2013) *Late Medieval Europe*, New York, NY: Routledge.

Watts, John (2009) *The Making of Politics*, 1300–1500, Cambridge: Cambridge University Press.

Weber, Max (1964) *The Theory of Social and Economic Organization*, New York, NY: The Free Press.

Wickham, Chris (2017) *Medieval Europe*, New Haven, CT: Yale University Press.

Wickham, Chris (2020) "Rethinking the Structure of the Early Medieval Economy," in Davis, Jennifer R., and Michael McCormick, eds. (2016) *The Long Morning of Medieval Europe*, New York, NY: Routledge.

Winroth, Anders (2004) *The Age of the Vikings*, Princeton, NJ: Princeton University Press.

6

STANDARDS AND EVOLUTION IN A COMPLEX WORLD

6.1 Introduction

Evolution is about survival and enhancement. It is the means by which actors/agents/genes/artifacts seek to augment their fitness through the process of emergence in the context of a changing environment. Central to this undertaking is a procedure whereby members of a population are selected to advance up a given hierarchy and be replicated at higher fitness levels in accordance with a set of fitness criteria—defined here as standards. Although present-day accounts of evolution incorporate many aspects of this characterization, a unified perspective, which encompasses all aspects of the world as we know it, has yet to be achieved[1] (Hodgson and Knudson, 2010; Turner and Maryanski, 2024, 4). Disagreements stem not only from disciplinary divisions but also from divergent methodological approaches, especially those regarding reductionism and the appropriate unit of analysis. However, as this chapter contends, various components of these diverse theories, when combined within the framework of complexity theory, can provide a more integrated, standards-based rendition of evolution.

6.2 Early Sociological Accounts

Consider, for instance, the elements of early sociological accounts. Emerging within the milieu of the 18th century Enlightenment, these efforts—which predated Darwin's writing of *The Origin of the Species*—sought to explain social change while maintaining a rigorous, positivistic approach.[2] Like the writers of the *Encyclopedia*, these burgeoning sociologists sought to identify the universal laws that govern social systems.[3] However, although referencing the material sciences, they viewed social evolution as a separate phenomenon, driven by its own unique forces (Turner and Machalek, 2018, 13–14). Many aspects of their works continue to be highly relevant to our present-day understanding of how complex systems emerge and evolve.

6.2.1 August Comte's Positivistic Sociology

August Comte (1798–1857), a philosopher and mathematician, is considered to have been the father of sociology. Building on Enlightenment efforts to employ the natural sciences for the

DOI: 10.4324/9781032721125-9

betterment of mankind, Comte successfully advanced the idea that the social world could only be understood scientifically through the lens of positivism (Bourdeau, 2023). In contrast to many Enlightenment authors who had focused on individual entities, Comte trained his analysis on society as a whole. As Maus (1998) opines:

> For Comte, society is a reality in itself, having a life of its own, which proceeds independently of all individual persons, and is not subject to their interference. Society is thus an organism in itself, and takes precedent over the individual. It is the real creator of civilization. (10)

According to Comte, to have an impact on social evolution required employing science to identify the laws governing it. Although Comte—like Darwin—was influenced by Malthus's projections about population pressures, he argued that each realm of inquiry was governed by its own set of laws, which could only be discovered through a specifically targeted, positivistic approach (Turner and Maryanski, 2004, 4–6). Hence, to preserve the discipline's integrity, Comte fiercely sought to maintain its independence from the natural sciences. In fact, in evaluating academic disciplines based on their scope and importance, Comte placed sociology at the very top of the list (Turner and Machalek, 2018, 17; Bourdeau, 2023). By preserving sociology's independence, Comte created an alternative path for analyzing evolution, one outside of, and beyond, the Darwinian model.

6.2.2 Durkheim's Social Facts and Niches

While Comte founded the discipline of sociology, it was Emile Durkheim who formalized the field, laying out its goals and methodology in *The Rules of Sociology,* published in 1895. Like Comte, Durkheim was a positivist who conceived of sociology as an independent discipline dedicated to explaining how social structures generate "social facts,"[4] which, he argued, determine social interactions and social transformations. Emphasizing the need to look beyond individuals, Durkheim claimed:

> If we begin with the individual in seeking to explain phenomena, we shall be able to say nothing of what takes place in the group.... We must ... seek the explanation of social life in society itself.
> *(Degler, 1991, 95)*

In accounting for social transformations, Durkheim drew upon the concept of natural selection popularized by Herbert Spencer and Darwin. However, because Durkheim's primary concern related to how societies maintain their coherence in the face of population pressures and the onset of modernity, his explanation of natural selection was far more subdued, and nuanced, than that proffered by Spencer (Turner and Machalek, 2018, 17). As posited in *The Division of Labor in Society* (1893), Durkheim viewed natural selection within and among societies as being generated by social differentiation—equivalent to biological speciation—and, with it, greater competition and more intense selection (Turner and Machalek, 2018, 13–14, 17). However, instead of fighting to the death for survival—as Herbert Spencer theorized—losers in a competition would, instead, withdraw from the battle so as to establish new social niches, along with the resources to sustain them, creating thereby a cyclical process entailing greater complexity; increases in the division of labor; and more intense competition for resources (Turner and Machalek, 2018, 17).

Durkheim reinforced the path set out by Comte. His emphasis on social groupings, and the social facts that govern their interactions, form a common thread that reappears throughout sociological accounts and explanations. These concepts are particularly relevant today in the context of complexity theory, where—appearing in the form of standards and hierarchical platforms—they serve as the building blocks of evolutionary changes.

6.2.3 Spencer's Organic Vision

It should, perhaps, come as no surprise that it was Herbert Spencer, a self-educated polymath, who first sought to integrate sociological and biological accounts of evolution. Active in the vibrant intellectual life of England and free from disciplinary constraints, Spencer was uniquely situated to seek out universal laws explaining evolution. Accordingly, anticipating Darwin,[5] and building upon Malthus, he proposed that, given continuing population pressures and a greater division of labor, societies could only survive by culling their weakest members and replacing them with superior ones, making way in the process for more complex, diverse structures that would ultimately be integrated—at the highest level—within a superstructure constituting society (Turner and Machalek, 2018, 26–27).

Although Spencer drew on Darwinism to account for evolution, he diverged from Darwin's paradigm in a number of significant ways (O'Connell and Ruse, 2021, 29). In contrast to Darwin, who, in his classic work, *The Evolution of the Species,* had characterized evolution as being relatively neutral in terms of its outcomes (Degler, 1991, 4), Spencer claimed that evolution was, in fact, purposeful, leading in stages from primitive societies to military societies, to industrial societies, and finally to self-regulated utopias. As he asserted,

> … whether the dangers to existence be of the kind produced by excess fertility, or of any other kind, it is clear, that by the ceaseless exercise of the faculties needed to contend with them, and by the death of all men who fail to contend with them successfully, there is ensured a constant progress towards a higher degree of skill, intelligence, and self-regulation—better coordination of actions—a more complete life.
>
> *(O'Connell and Ruse, 2021, 29–30)*

Spencer also differed from Darwin in characterizing the pace, as well as the course, of evolutionary change. Whereas Darwin conceived of evolution as occurring incrementally, over extended periods of time, Spencer built on the Lamarckian concept of intergenerational inheritances, whereby offspring might acquire traits developed during the lifespan of their parents. As importantly, contrary to Darwin's view that individual entities formed the fundamental units of the evolutionary process, Spencer, much like Durkheim, emphasized that individual actions were, themselves, structured in the course of evolution, due to their integration in collective social entities, which—arranged hierarchically—extended from individuals to groups to societies at large (O'Connell and Ruse, 2021, 107). Thus, Spencer contended that evolution took place not, as Darwinians argued, in a continual, ongoing process, but rather in stages, whereby individuals, embedded in social structures, emerge from one level of the evolutionary hierarchy to the next.

While agreeing with Darwin that the battle for fitness—and hence for survival—was the primary mechanism fueling biological evolution, Spencer extended his analysis to the social arena, claiming that, much like all organic entities, social groups and societies themselves competed in an evolutionary battle for survival of the fittest (Hodgson and Knudson, 2010, 7–80).

As Spencer emphasized, the battle for survival was a natural, and hence, unalterable given. Thus, the only option for society was to make way for it, or—as Social Darwinists proposed—to promote its progressive results through regressive social engineering, including eugenics (O'Connell and Ruse, 2021, 30–32).

To account for evolutionary social outcomes, Spencer introduced a functionalist element into his analysis. Viewing societies as organic systems, he claimed that, much like natural systems, in which traits are selected to support the whole, so too social evolution must select for interrelated structures that meet system-wide criteria (O'Connell and Ruse, 2021, 51–52). According to Spencer, these prerequisites included mechanisms for carrying out production; means to ensure the reproduction of individuals and social units; systems for regulating and consolidating power; and infrastructure for distributing information and resources (Turner and Machalek, 2018, 26–27).

As Richard Hofstadter opined, Spencer "was … the profoundest thinker of [his] time" (Hofstadter,1955, 63). So compelling were his ideas that, by the first quarter of the 19th century, his works had been translated throughout Europe. But it was in the United States, where Spencer had his greatest impact. As Hofstadter (1955) relates:

> Spencer's philosophy was admirably suited to the American scene. It was scientific in its derivation and comprehensive in its scope. It had a reassuring theory of progress based upon biology and physics. It was large enough to be all things to all men, and broad enough to satisfy agnostics … It offered a world- view, uniting under one generalization everything in nature from protozoa to politics. Satisfying the desire of "advanced thinkers" for a world-system to replace the shattered Mosaic cosmogony, it soon gave Spender a public influence that transcended Darwin's … It made Spencer the homemade intellectual, and the prophet of the cracker-barrel agnostic. (63–64)

Notwithstanding Spencer's extensive popularity, as the 19th century wore on, people became repulsed by the eugenic policies of Social Darwinism, as well as increasingly skeptical about his theoretical predictions (Hodgson and Knudson, 2010, 16–17).[6] Instead of the promised utopia, the new century witnessed not only World War One and the Great Depression, but also—in the aftermath of these events—a call for greater social collaboration and the rise of socialism (Larson, 2004, 211–216; O'Connell and Ruse, 2021). Nevertheless, two major strands of Spencer's thought have endured, reappearing today in the context of complexity theory. The first is Spencer's (dare I say) functionalist conceptualization of society as an interdependent organic entity, whereby the whole is greater, and more complex, than the sum of its parts. The second is his proposition that evolution is purposeful, entailing discreet stages, each of which emerges from the previous one at a higher level of fitness.

The decline of Social Darwinism led to the retreat of sociologists from the pursuit of evolutionary theories. As Hodgson and Knudson point out, "For many, because of its biological associations, the very word *evolution* became taboo." Characterizing the situation, these authors note: "It was the beginning of the dark ages for evolutionism in the social sciences" (Hodgson and Knudson, 2010, 16).

6.3   The Evolution of Evolutionary Science

Spencer, Comte, and Durkheim were not alone in their exhortations for a scientific understanding of evolution. Early efforts to promote positivism and rational thinking had emerged as part of the 17th century "scientific revolution." Building on these scientific undertakings, philosophers and scientists in the Age of Enlightenment sought to compile this entire body of knowledge, systematically categorizing and interpreting it within a unified, scientific compendium, culminating in the *Encyclopedia*. Mirroring the composition of the intellectual elite, these philosophers and scientists were highly diverse both with respect to their backgrounds and expertise. Thus, while sharing the common goal of promoting science, they often disagreed on issues, especially those relating to theological questions and prevailing doctrines and institutions (Gay, 1967).

Because the *Encyclopedia* and the issues it posed were discussed widely among the literate public in academies, learned societies, coffee houses, and public lectures, discussions and debates about evolution and theology quickly became popularized, creating the contentious environment in which evolutionary theory emerged (Morley, 2002). As Larson (2004) describes:

> Some naturalists and savants challenged static concepts in science, including the fixity of the species; many rejected any role for the supernatural in the natural. Rational materialism gained ground in scientific, social, and political thought—with no clear separation among these disciplines. Disorder became the order of the day, and a reaction became inevitable. (33)

As we shall see, resolving these issues was a lengthy process, requiring the aggregation of ever more extensive and conclusive evidence along with the scientific tools to evaluate it.

6.3.1 The Creationist Challenge

Just as sociologists predated Darwin in their efforts to account for evolutionary changes, so too did the philosophers and naturalists of the day. However, their efforts were hindered to a considerable degree by the prevailing belief in the biblical account of creation found in the book of Genesis. For, if God had created life at one stoke, and in a perfected form, as Genesis proclaimed, evolution would be precluded by definition (Larson, 2004, 38–41). This theological worldview of a fixed and perfect universe persisted even among scientists, notwithstanding the growing fossil evidence demonstrating an organic succession from the simplest forms of life at the bottom of geological layers to more complex ones at the top[7] (Larson, 2004, 44–57). To explain this anomaly, while maintaining the basic principles of natural theology, naturalists and philosophers posited that God created a succession of species, each of which was specifically designed to inhabit a unique geological epoch, in keeping with his overall plan for Earth (Larson, 2004, 53–54).

6.3.2 The Biological Account

Darwin's work, *On The Origin of the Species,* challenged these ideas. His conclusions were developed during his voyage on the Beagle, which was for him a transformative experience. Prior to this trip, Darwin had been a firm believer in the theory of catastrophism, which held that evolution took place in a discontinuous manner, allowing God to create new populations best suited to their altered environment, while eliminating those no longer sustainable within it (Larson, 2004, 57–58). Darwin's observations in the Galapagos altered his perspective: upon his return, he renounced catastrophism in favor of Charles Lyell's alternative theory, which posited that long-term geological changes were gradual and cyclical, rather than abrupt and directional.[8] Based on his observations of the biological differences between finches, and between those of mockingbirds, Darwin claimed that distinctions between species were due to their adaptations within separate niches, a finding that he claimed accounted for the origin, and variety, of diverse species.[9]

Notwithstanding Darwin's findings, and despite mounting fossil evidence, his contemporaries remained unconvinced.[10] (Larson, 2004, 68–69). To counter this opposition, Darwin had to further theorize the process of evolution and the means by which it took place. His reading of Malthus's treatise on population growth provided a compelling explanation of the struggle for survival[11] (Larson, 2004, 87–89). According to Malthus's *Essay on the Principle of Population* (1789), populations were destined to increase at an exponential rate, exceeding the growth of the food supply, leading to famines and warfare, and thereby reducing populations to sustainable levels. Incorporating Malthus's thesis into his own framework, Darwin contended that the evolution of species followed a similar pattern: just as growing populations declined in the face of competition for resources, so too an increase in the number of species would rise and fall in a battle for survival—or, as he later characterized the process, "the survival of the fittest."[12] Moreover, Darwin drew upon Adam Smith's conception of the *invisible hand* to account for how the individual interactions at a local level might, over time—and albeit slowly—give rise to innovative global outcomes. It was upon this basis that he claimed that natural selection constitutes a creative force (Gould, 2002, 105–106, 116, 123).

Darwinians also had to identify how individual traits were passed on from one generation to the next (Larson, 2004, 148–149; Kirschner and Gerhart, 2005). To explain this piece of the puzzle, they turned to Mendel's re-discovered, statistically-based theory of mutation, which Mendel had derived from his research on the crossbreeding of thousands of pea plants. With this new information, Darwinians could explain variation by virtue of a hereditary process. Nevertheless, a general agreement about how reproduction, inheritance, and variation operated was not achieved until the turn of the 20th century, when Watson and Crick identified the gene's double-helic structure, whereby scientists came to understand that hereditary information was carried on chromosomes in discreet units called genes (Larson, 2014, 183, 241). This discovery, together with the biometric research of Haldane, Fisher, and Sewall Wright, regarding the effects of genetic changes on populations, opened up a new field of "population genetics," and along with it a reductionist account of evolution (Larson, 2014, 241–242).

6.3.3 The Evolutionary Synthesis

Throughout the forties and fifties biologists sought to develop a consensus about how evolution operated. In the first phase, they banned Lamarckian thinking,[13] and coalesced around a new theory of Mendelism based on population genetics. Other more contentious issues, such as saltation (major shifts in evolutionary processes) and orthogenesis (purposeful evolutionary outcomes), which were considered to be of minor significance, were of put aside for future resolution[14] (Gould, 2002, 383). But such a resolution was never forthcoming. It was only in 1959, during the Darwinian centennial celebration, that a final synthesis was achieved. By then, evolutionary biologists' opinions had not only narrowed; the process of achieving agreement had become far more constrained. The new synthesis, referred to as neo-Darwinism, characterized natural selection as the exclusive agent of evolutionary change, ruling out, among other things, the possibility of group selection, staged evolution, and adaptation to environmental conditions (Gould, 2002, 384–386).

6.3.4 Push-Back against the Consensus

Failing to provide a complete picture of how evolution actually functions, the Modern Synthesis left much open to debate. As Kirschner and Gerhart (2005) describe the results:

> The Modern Synthesis reduced evolution to three basic steps. First, there was the occurrence of random genotypic variation—in modern parlance, a random modification of the sequence of DNA. Second, the change of genotype caused a change of phenotype within the individual organism (by means not specified). Third, the altered phenotype was selected (and with it the altered genotype required for it) on the basis of the individual's reproductive fitness, that is its ability to contribute progeny to future generations. On the question of how the altered genotype caused an altered phenotype, the Modern Synthesis was silent. (45)

Given this outcome, we can see why, over time, and in the face of new evidence and theoretical claims, a number of alternative perspectives have emerged. In the 1970s, for example, Niles Eldredge and Stephen J. Gould contended that the course of evolution is not uniform as neo-Darwinians claimed; instead, it is subject to long periods of stasis, punctuated by sudden bursts of innovation[15] (Gould and Eldredge, 1977). In turn, John H. Mayfield introduced the idea that evolution is a process based on computation and the underlying encoded instructions that direct it.[16] Based on this assumption, Mayfield claims that—contrary to neo-Darwinian perspective—evolution must be purposeful given that instructions always have intended outcomes (Mayfield, 2013, 31).

Others, while adopting the Darwinian narrative, have pointed out gaps in its logic. They call for more specific information about how evolution works. For one, they want to know whether mutations, which are said to emerge through a lengthy, step-by-step, process, can produce a sufficient number, and adequate variety, of genotypes to generate novel, complex phenotypes.[17] (see Behe, 2002; Kirschner and Gerhart, 2005, 9–10, 20–25). Questions have also emerged about how evolutionary changes, originating in genotypes, can be translated into changes in phenotypes. For, according to the Modern Synthesis, the variation of genotypes is totally uncoupled from selection among phenotypes, even though it is by virtue of the phenotype, and its offspring, that revolutionary outcomes are ultimately realized. Still to be explained, moreover, is the question of how complex entities, comprised of a number of distinct, functional parts, are assembled, absent some intervening interactions, in a linear process. One must ask, for example, where, when, and by what means are the necessary interconnections and agglomerations put into force.

Efforts to resolve these conundrums, based on new insights and analyses, have generated a number of hypotheses: but, many of these evoke additional uncertainty about some of the fundamental tenants of the neo-Darwinian account. Consider for instance, Kirschner's and Gerhart's concept of conserved, core cell constructs, which serve much like Lego Blocks, allowing cells to be easily reassembled, reconfigured, and reused in creative ways. As the authors argue, because they bypass the need for drastic cell modifications and cell innovations, these core processes greatly reduce the time required for evolutionary changes to take place (Kirschner and Gerhart, 2005, 55). Note, however, that while this conceptualization might account for the speed of evolutionary changes; the efficient use of genetic materials; as well as the means by which organisms come together to form genetic structures, it belies the

Darwinian assertation that changes take place slowly, in a fluid, linear fashion. As Kirschner and Gerhart (2005), note:

> Darwin's supposition that change is pervasive has to be replaced with the view that in the history of life some things change, and others do not, and that change occurs in spurts and then become fixed and subsumed in all descendant organisms…. (85–86)

Other scholars have sought answers outside the neo-Darwinian paradigm. Complexity biologist Stuart Kauffman, for example, has made the case that evolution is not the product of selection alone, but also of a spontaneous ordering of self-organized systems, referred to as self-organizing catalytic sets (Kauffman, 1995, 115, 39; see also Kauffman, 1993, 2019, 2011). As he professes:

> Life, in this view, is an emergent phenomenon arising as the molecular diversity of a prebiotic chemical system increases beyond a threshold of complexity. If true, then life is not located in the property of any single molecule—in the details—but is a collective property of systems of interacting molecules …. Life, in this view, is not to be located in its parts, but in the collective emergent properties of the whole they create … the collective system does possess a stunning property not possessed by any of its parts. It is able to reproduce itself and evolve. The collective system is alive. Its parts are just chemicals.
>
> *(Kauffman, 1995, 38)*

While Kauffman's conceptualization accounts for how cells come together in a self- organizing phase transition, to form more complex cellular structures, it also reinforces Eldredge's and Gould's arguments with respect to stages and punctuated disequilibrium.

6.4   The Sociological Response

At this point, we must wonder: were Comte and Durkheim correct in arguing for a clear distinction between the biological and the sociological accounts of evolution? It is a posture that many social scientists have adhered to, up until the present. In fact, it is the reason why—given the decline of Talcott Parson's theory of structural functionalism,[18] most sociologists have shied away from the subject of evolution altogether (Hodgson and Knudson, 2010, 8). Their reluctance to tackle evolutionary issues has only been buttressed by the works of E.O. Wilson, which not only established the field of sociobiology, but also, with its focus on genetics and its determinist undertones, led to a reductionist shift in evolutionary thinking.[19] The field rapidly gained momentum following evolutionary biologist Richard Dawkins's writings[20] claiming that the gene, not the individual, is the ultimate unit of selection—or more precisely, it is where the development and behavior of individuals and behavior are decided. (Turner and Machalek, 2018, 111).

By definition,[21] a reductionist approach eliminates many of the issues of concern not only to early sociologists, but also to those dissident biologists who have questioned the Evolutionary Synthesis. From their perspectives, still to be explained are the social units of replication and selection; how they are interrelated; where, how, and at what stages in the process do they interact; as well as, to what extent, and by virtue of what means, is agency entailed in their execution (Turner and Machalek, 2018). These theoretical omissions have made it difficult for those few sociologists who have ventured forth to find common ground with Neo-Darwinists. At the same time, sociobiologists—intent on their genetic focus—have not yet made a meaningful effort to look for insights within sociological paradigms (Turner and Machalek, 2018). The result, we might say, is cross talk.

6.5 Complexity—A Unifying Theory?

Can complexity theory serve to accommodate—if not link—these diverse viewpoints? One might think so. For the phenomenon of emergence is, as we have seen, central to the theory of complexity. In fact, we might say that emergence is, itself, an evolutionary process. Thus, it should come as no surprise that emergence, as characterized in complexity theory, is in several ways congruent with both the sociobiological and the sociological perspectives.[22]

Consider first the case of sociobiology. Like sociobiology, complexity theory views interactions, and the standards that govern them, as the mechanisms driving evolutionary change. Moreover, both analytical frameworks assume that selection, operating in accordance with some fitness criteria—whether it be reproductive capacity or adaptability—is essential for evolution to take place. Furthermore, notwithstanding disagreements as to whether evolutionary processes are linear or complex, both narratives posit that change occur in an upward direction leading to greater levels of complexity. They agree, moreover, that, evolutionary outcomes are based on standards, taking the forms, such as codes.[23] As complexity theorist Stewart Kauffman tell us in this book *At Home in the Universe* (1995), it is only by virtue of the standard interfaces inherent in the universe that diverse entities that comprise all phenomena were able to interact, repair, coevolve, and recreate themselves. According to Kauffman, it is the laws of the universe, embodied in these standards, that have given rise to a natural, hidden order. These consequences are unpredictable. Where complexity theory diverges from sociobiology is at the point where the focus of analysis shifts from the genotype to the phenotype—the area most amenable, and thus most of interest, to sociologists and other behavioralists.

Often, it is said that, while genetic interactions allow for evolutionary changes, in the long run, they cannot fully account for them. To understand how, and why outcomes end up ultimately as they do, we need to home in on the phenotype. Incorporating many of the insights from evolutionary sociology, complexity theory offers a framework for accomplishing this. As detailed in the previous chapters, complexity theory posits that evolutionary changes occur in stages, atop of platforms, which are layered hierarchically, in accordance with their levels of fitness and complexity. Each platform is the site of interactions, which are governed by the platform's unique set of standards, which are determined and updated during the course of interactions. Platforms are also linked to their external environments from where they import and exchange new information and resources. By agglomerating these external inputs together with their own internal resources, platforms generate new, more adaptive standards, which allow platform actors or agents—as the case may be—to emerge higher up on the fitness landscape.[24] As Mayfield has noted, these adaptive standards, or computational rules as he describes them, serve as the "engine of complexity" (Mayfield, 2013, 20–21).

6.6 Revisiting Durkheim, and His Concept of Niches

There is much to be gained by narrowing the gap between gene-based biology and agent-based sociology, so as to allow us to view life as the product of unified forces. By doing so, we may advance our understanding of many puzzles, which have hitherto defied resolution. Recent efforts to capture the life of plants provide a prime example. Most notable in this regard is the recent work of Zoe Schlanger, *The Light Eaters: How the Unseen World of Plant Intelligence Offers a New Understanding of Life on Earth* (2024). While building upon Darwin's insights, Schlanger discovers amazing answers and fundamentally new questions when turning her attention from genes to plant behavior.[25]

Consider some of Schlager's findings and how she came to account for them. As we have seen, a challenge for sociobiologists has been to explain how individual cells form complex, multi-functional organisms. To do so, Schlager argues that these cells not only need to communicate and coordinate their behavior with other cells, but they must also understand their own role in the chain of events, which they come to grasp through their interactions with other plants. As she describes, cells perform this function on what complexity theory might describe as *platforms*. Referencing a 2017 study at the University of Birmingham that has identified "decision-making centers" within dormant seeds, she notes that cells employ these sites to share information about existing conditions, in order to determine when the plants should emerge. As she points out, and in keeping with Stewart Kauffman's claims with respect to autocatalytic sets, it is through the chatter of the cells that plants self-organize (Schlanger, 2024).

Plants also are intentional in using their communication skills and chemical resources to enhance their own wellbeing by engaging with the members of their environments. In doing so, they are able to differentiate among plants in their environment. Researchers have found, for example, that sagebrush plants are more responsive to cues signaling danger when they are sent from a close family member. Similarly, goldenrods behavior is related to the state of their environment. Where predators are few, goldenrods will, when under attack, send out warning signals only to their close kin, whereas, where predation is more serious, they will broadcast their signals throughout their surrounding area (Schlanger, 2024). As Schlager (2024) suggests, this behavior

> makes sense … in terms of community survival; when the whole neighborhood is threatened, it is best to save as many as of your kind as possible, regardless of if they are family or not.… This means plants could be said to have dialects, and they are alert to their contexts enough to know when to deploy them. More than that, they have a clear sense of who is who, who is family, and who is not. They are in touch with their surroundings, and with the fluctuating status of their enemies. Their communication is not just rudimentary but complex and layered, alive with multiple meanings.

Schlanger's work is not only tremendously insightful; it also illustrates the enormous value of viewing evolution not only from a genetic perspective, but from a behavior perspective, at the level of the phenotypes, as well. Much like Durkheim's idea of niches, which interact to enhance their resources, while competing with other niches; as well as complexity's focus on interactive platforms, which adapt and emerge in stages through their interactions with their environments; Schlanger's ecological investigation of plant communities—which are themselves complex adaptive systems—provides the means of viewing evolution, not by its individual parts, but rather as a whole.

Notes

1 As Hodgson and Knudson point out: "… an adequate refinement of general Darwinian concepts such as selection, replication, and inheritance in terms that could be applied to social or economic."

2 As Carl Degler notes, "Although some ancient Greek thinkers had recognized change as inherent in life, few in antiquity and virtually none in early Christian Europe took seriously the idea that the present had emerged out of the past.. . . . Immanuel Kant's assertion that the universe is the product of slow changes over eons of time was amongst the earliest examples of an evolutionary outlook" (Degler, 1991, 3–4).

3 As Turner and Maryanski point out, "early sociologists such as August Comte, Herbert Spencer, and Emile Durkheim were all seeking science with precision and abstract theory like that of physicists, with Sir Isaac Newton's 'laws of gravity' being the ideal for social science, and theorizing on the evolution of socio-cultural formations from simple to more complex forms" (Turner and Maryanski, 2024).

4 Emile Durkheim employed the term "social facts" to describe social structures that set the parameters on individual and group behaviors. For a discussion, see Elder-Vass (2014).

5 As Turner and Maryanski point out: "Spencer did not use Darwin's notion of natural selection because, in his mind, he had already formed this idea in philosophical works a decade before Darwin's publication of *On the Origin of the Species by Means of Natural Selection* in 1859" (Turner and Maryanski, 2004, 8).

6 As Hodgson and Knudson described: "In the surge of nationalism before and during the First World War, phrases such as. Spencer's *survival of the fittest* and Darwin's *struggle for existence* were given nationalist and racist associations. Vaguely Darwinian ideas were also bandied about to justify or illustrate all sorts of contradictory social and political stances, including nationalism, militarism, imperialism, free trade, individualism, socialism, and even pacifism" (Hodgson and Knudson, 2010, 16).

7 For example, with the discovery of dinosaur fossils in the mid-17th century, European naturalists, such as William Buckland, could begin to see a progressive direction in the record. Similarly, geologist Adam Sedgwick, who discovered the Cambrian System, found that, although the period evidenced sharp breaks in the fossil record, it suggested at one and the same time a progressive trend line (Larson, 2004, 52–57).

8 As Gould claims: Natural selection, according to Darwin, "relies entirely on small isotropic, non-directional variation as raw material and views extensive transformation as the accumulation of tiny changes wrought by struggle between organisms and their (largely biotic) environment. Trial and error, one step at a time, becomes the central metaphor of Darwinism" (Gould, 2002, 85–86).

9 As Larson relates: "For Darwin, a species simply constituted a population of physically similar individuals capable of breeding together, not an ideal, unchanging life form. … He realized that different natural environments would favor different adaptions so that species would not evolve in a linear, Lamarckian fashion. Rather, Darwin envisioned a branching process of evolution, with various daughter species evolving in different directions from a common ancestral type to fill available geographic spaces and biological niches. For Darwin, different death rates caused by natural factors created new species. God was superfluous to the process" (Larson, 2004, 89).

10 For example, as Stephen Jay Gould pointed out, although many anti-Darwinian biologists agreed upon the notion of natural selection as a causal force in nature, they refuted the idea that it resulted from a battle for survival (Gould, 2002, 28–29).

11 Note that Comte, Durkheim, and Spencer also drew upon Malthus to develop their theories.

12 It was Herbert Spencer who coined the term "survival of the fittest and suggested that Darwin employ it in referring to a battle for survival.

13 Lamarckian theory of heredity posits that organisms can transfer physical traits to their offspring which they have acquired through use or disuse during their lifetimes.

14 As Gould has opined: With the coalescence and hardening of the Modern Synthesis culminating in the Darwinian celebrations of 1959, an orthodoxy descended over evolutionary theory, and a generation of unprecedented agreement ensued (often for reasons of complacency or authority). However, the press of new concepts and discoveries has since fractured this shaky consensus, and we now face a range of options and alternatives as broad as previously (Gould, 2002, 134).

15 As the authors contended: "We substitute a hierarchical selectionist theory of numerous interaction levels, a balanced and bidirectional flow of causality between external selection and internal constraint (interaction of functionalist and structuralist perspective), and causal interactions among tiers of time" (Gould, 2002, 42).

16 As Mayfield argues, evolution is a process based on information accumulation. It takes place by pursuing a particular strategy for information manipulation and accumulation. This process is one of computation (Mayfield, 19–24).

17 Marc W. Kirschner's and John C. Gerhart's evolutionary account in their book *The Plausibility of Life: Resolving Darwin's Dilemma* illustrates why this might be the case. As the authors point out, evolutionary selection always acts on the phenotype and only indirectly on the genotype *via* the phenotype. Thus, if selection acts as a sieve preserving some variants but not others, one must then wonder whether, if genetic variation is random, as Darwinists claim, enough phenotypic variation will take place to allow for successful adaptations. Moreover, because genes are also uncoupled from the environment, evolutionary adaptation occurs by virtue of changes in the phenotype.

18 See for a discussion of Parson's social theory, Alexander, Jeffrey (1983).

19 A key turning point was the publication of George C. Williams's book, *Adaptation and Natural Selection: A Critique of Some Current Evolutionary Thought* (1966), in which the author challenged the sociological contention that selection produces traits on behalf of the group while contending, instead, that selection operates to enhance the benefit of the individual.

20 See, for instance, Richard Dawkins (2016).

21 As argued by Bossomaier and Green, "Adhering to a reductionist analytical framework, neo-Darwinians and social biologists conceive of evolution—by definition—as a deterministic, linear process driven by alterations in individual genes. Left unaccountable in their account is a satisfying explanation of how individual genes are reconfigured as increasingly complex structures, as well as what role that individual agency and environmental factors play in their construction" (Bossomaier and Green, 1998).

22 Complexity biologist Stewart has taken major steps in this regard by characterizing cells as complex adaptive systems, which evolve based on rules that have evolved, from the bottom up, in reaction to the actions of their component parts, as well as to their changing environment (see Kauffman, 1995).

23 Describing the role of standards in cellular behavior, Goodsell notes that, like the standardized interchangeable parts of modern machines, the components of molecular machines connect with each other when their parts—defined by their chemical make-up—fit snugly together. Although molecules encounter one another randomly when swimming in the cell's fluid environment, they only bind together when the interaction is complementary—that is to say, when their interfaces are perfectly matched to a common standard (Goodsell, 2009).

24 For a detailed, empirical account of how this process takes place, see Chapter 2, "Assent Up the Fitness Landscape: How the West Was Won."

25 Explaining her point of view, Schlanger notes: "It is no wonder then that zoologists and entomologists have been the ones to make some of the most groundbreaking discoveries about plants, often by viewing them through the lens of animals and insects. That's not to knock botanists, but in an age where genetics dominate, many have ceased to see the plant as a pulsating whole and instead see it as an amalgam of genetic switches and protein gates (Schlanger, 2024).

References

Alexander, Jeffrey C. (1983) *Modern Reconstruction of Classical Thought: Talcott Parsons*, Berkeley, CA, California University Press.

Behe, Michael J. (2002) *Darwin's Black Box: The Biochemical Challenge to Evolution*, New York, NY, The Free Press.

Bossomaier, Terry, and David Green (1998) *Patterns in the Sand: Computers, Complexity, and Everyday Life*, Reading, MA, Persius Books.

Bourdeau, Michel (2023) "August Comte," *The Stanford Encyclopedia of Philosophy*, https://plato.stanford.edu/archives/spr2023/entries/comte/

Dawkins, Richard (2016) *The Selfish Gene*, 4th edition, New York, NY, Oxford University Press.

Degler, Carl N. (1991) *In Search of Human Nature: The Decline and Revival of Darwinism in American Social Thought*, New York, NY, Oxford University Press.

Elder-Vass, David (2014) *The Causal Powers of Social Structures: Emergence, Structure, and Agency*, New York, NY, Cambridge University Press.

Gay, Peter (1967) *The Enlightenment: An Interpretation*, New York, NY, Alfred A, Knopf.

Goodsell, David S. (2009) *The Machinery of Life*, 2nd edition, New York, NY, Springer Science & Business Media.

Gould, Stephen J. (2002) *The Structure of Evolutionary Theory*, Cambridge, MA, Harvard University Press.

Gould, Stephen J., and Neils Eldredge (1977) "Punctuated Equilibrium: The Tempo and Mode of Evolution Reconsidered," *Paleobiology* 3 (2), 115–151.

Hofstadter, Richard (1955) *Social Darwinism in American Thought, With a New Introduction by Eric Foner*, Boston, MA, Beacon Press.

Kauffman, Stuart (1993) *The Origins of Order; Self-Organization and Selection in Evolution*, New York, NY, Oxford University Press.

Kauffman, Stuart (1995) *At Home in the Universe: The Search for the Laws of Self-Organization and Complexity*, New York, NY, Oxford University Press.

Kauffman, Stuart (2011) "Approaches to the Origin of Life on Earth," *Pub Med Central* 1 (1), 34–48, https://doi.org/10.3390/life1010034

Kauffman, Stuart (2019) *The World Beyond Physics: The Emergence and Evolution of Life*, New York, NY, Oxford University of Life.

Kirschner, Mark W., and John C. Gerhart (2005) *Resolving Darwin's Dilemma*, New Haven, CT, Yale University Press.

Larson, Edward J. (2004) *Evolution: The Remarkable History of a Scientific Theory*, New York, NY, Random House.

Maus, Heinz (1998) *A Short History of Sociology*, New York, NY, Routledge.

Mayfield, John E. (2013) *The Engine of Complexity: Evolution as Computation*, New York, NY, Columbia University Press.

Morley, John (2002) *Diderot and the Encyclopedists*, Chapman Hall, Open Library.

O'Connell, Jeffrey, and Michael Ruse (2021) *Social Darwinism*, New York, NY, Cambridge University Press.

Schlanger, Zoe (2024) *The Light Eaters: How the Unseen World of Plant Intelligence Offers a New Understanding of Life on Earth*, New York, NY, Collins.

Turner, Jonathan H., and Alexandra Maryanski (2024) *Biosocial Analysis, Reconciling Biology, Psychology, and Sociology*, Cheltenham, UK, Edward Elgar Publishing Limited.

Turner, Jonathan H., and Richard S. Machalek (2018) *The New Evolutionary Sociology: Recent and Revitalized Theoretical and Methodological Approaches*, New York, NY, Routledge.

PART III

Standards

The Building Blocks of Complex Outcomes

7

COMPLEX MONETARY OUTCOMES— WINNERS, LOSERS, AND THE STANDARDS DETERMINING THEM

7.1 Introduction

Ten years ago, protesters converged at Zuccotti Park on Wall Street to express their opposition to the structure and performance of American capitalism. Their slogan, "We are the 99 percent," sought to make the public aware of the tremendous discrepancies in income distribution in the United States, the fact of which few Americans were fully aware. Blaming Wall Street and the Great Recession, the protesters claimed that there was something terribly wrong with the U.S. financial system. However, journalists and other pundits were skeptical of the Occupy Wall Street event, and they downplayed its importance, citing a lack of leadership and a clear strategy and structure.

While the ambiguity of the Occupy Wall Street event allowed the movement to bring together a wide array of participants with different, albeit congruent, concerns, its nebulous approach failed to relate the problem of a dysfunctional financial system to the negative effects of large-scale poverty and inequality (Dodd, 2014). If we want to rethink how our financial system should function, we need to better understand money as a standard and how it accounts for major, emergent outcomes Complexity theory provides a clue.

In this chapter, I argue that money is not neutral, as many economists have contended. To the contrary, serving as a standard, money generates the emergence of winners and losers, as is the case with all standards (Holland, 1995). However, to the extent that money operates in a complex environment, the outcomes of monetary interactions are far from predictable. That being the case begs the question of how we are to evaluate or guide our monetary policies. As importantly, if the economy falls short, who and what is to blame? This was the question that preoccupied the protesters and pundits focusing on Occupy Wall Street.

Stepping back from the specifics of the situation, we can employ complexity theory to help us not only identify the role that money plays in distributing wealth but also determine how it generates long-term social structures. As we shall see, comprised of standards, the institutional platform from which money emerges is a critical factor in determining monetary outcomes.

DOI: 10.4324/9781032721125-11

7.2 Taking Money for Granted

As I encountered it as a four-year-old, money is both awesome and perplexing. From my home on East 23rd Street, in Paterson, New Jersey, I could smell the enticing aroma of the Wonder Bread factory, where white bread, cream-filled chocolate cupcakes, and other sweet delights were made. My great-grandfather, Grandaddy Dillistin, would stop there on his way home from his lumber store to select a treat for my sisters and me. But that was not the only place to buy such delicacies. Around the corner from my house was a store where one could purchase newspapers, comic books, soda pops, and—of course—penny candy. It was here that I ventured out one evening in my pajamas to seek such sweet delights. My favorites were the different colored sugar dots lined up in rows and columns on long strips of paper. As I nibbled on the dots, making new patterns with each bite, I marveled at the treasure a penny could buy. Clenching my pennies in my fist, I was in wonder of these little copper coins.

How things have changed! Remember Benjamin Franklin's adage in his book, *Poor Richard's Almanac*: "A penny saved is a penny earned." Yet today's merchants are often willing to round out a bill, just to avoid dealing with pennies. And recall the saying, "Find a penny, pick it up, and all the day you'll have good luck." Despite the promise, only rarely do people stoop to pick up a penny. What happened to the penny's value? As Galbraith asked: Where did it come from; where did it go (Galbraith, 1995)?

Like most people, the more I dealt with money, the more I took it for granted. However, in my first economics course at Syracuse University, I learned, much to my surprise, that money was not backed by an equivalent amount of gold. I was not alone in my amazement. Arriving at my sorority house for lunch, I shared my discovery with my sorority sisters and our housemother, Mrs. Franz. My sorority sisters were curious and asked for more detail. But my housemother was aghast! "Girls, girls, girls," she said. Patting her hand rapidly on her chest, as she was wont to do when unnerved, she declared our luncheon over. As I made my way back to my room, I knew that, if money could arouse such agitation, it must be a subject to pursue.

7.3 What Is Money? Where Does It Come from?

Money continues to be bewildering, even as it has captured our fascination for centuries. Is money—as some have claimed—a thing, a veil, a process, a number, a social technology, or a medium of exchange? Absent some general agreement on the nature of money and how it operates, we cannot employ our monetary policies to their best advantage (Hodgson, 2015, 32–60). Instead, given our competing and often contentious frames of reference, our policy prescriptions are typically intertwined with ideological biases and power considerations. The outcome has been far from optimal.

The need to clarify our understanding of money has become even more salient due to a number of circumstances and events (Davis, 2010; Shiller, 2012). The recession of 2008 raised questions, for example, about the viability of a financial system that sought to reduce debt by incurring more debt (Dodd, 2014, 11; Martin, 2015, viii). As importantly, the Occupy Wall Street Movement focused attention on the role banks play with respect to the growing gap between the rich and the poor (Dodd, 2014, 24–25). At the same time, the rise of cryptocurrency has generated a growing interest in decentralized money, while simultaneously setting off alarms about the impersonal nature of electronic currency, as well as the loss of government control. Most recently, the coronavirus pandemic, the ensuing economic inflation, as well as the recent banking crisis, have brought such concerns to the fore.

In this chapter, I conceptualize money abstractly, defining it as *a standard of value*. Such a definition conceives of money not only as a standard that mediates interactions but also as one that, in the process, generates emergence and complex outcomes. By comparing money and its uses over time and in different contexts, we can see how, as societies have become more complex (as measured, for example, by the number and heterogeneity of the actors, their decentralization, as well as their various norms, standards, and formal laws), monetary ramifications have become more variable and uncertain. If, as some have argued, we are on the verge of a major transformation of our monetary system, we should, at the very least, know what we are dealing with. In this chapter, I argue that complexity theory can provide a guide.

7.4 Conceptual Controversies

Just as money has evolved over time, so too have our ideas about money. As conceived by Aristotle and then again, many years later, reiterated by Adam Smith, and later Carl Menger, scholars have tended to view money—whether in the form of tokens, shells, coins, etc.—in terms of its convenience for market exchange (Smithin, 2004; Martin, 2015; Orrell and Chlupaty, 2016, 21–22). This way of thinking attributes the emergence of coinage—that is to say, commodity money—to interactions in the limited sphere of the marketplace. Despite a lack of historical evidence, this perspective has persisted among mainstream economists until the present (Ingham, 2004; 15–16; Peacock, 2013, 30–31; Martin, 2015, 10; Orrell and Chlupaty, 2016, 30).

7.4.1 The Tenacious, Neoclassical Perspective

I first came across the neoclassical perspective years ago when studying economics based on Paul Samuelson's classic volume, *Economics: An Introductory Analysis* (1973), originally published in 1948. In my day, this textbook served as the font of all undergraduate economic wisdom. As in most neoclassical analyses, Samuelson employed a functional argument based on methodological individualism. Accordingly, he posited that individuals, interacting in the market and pursuing their own self-interest, logically progressed from bartering goods and services to a trading system based on coinage—that is to say, commodity money pegged to a common standard. Samuelson contended that this evolutionary path emerged spontaneously, driven by the desire for convenience, greater efficiency, and a reduction in transaction costs. Moreover, Samuelson's scenario relies, in some form or another, on self-organization, or—one might say—the invisible hand (Orrell and Chlupaty, 2016, 14). In this respect, it seeks to explain the past in terms of today's institutional structures—somewhat of a problem, as we shall see (Peacock, 2013, 19, 30–31).

Today, most neoclassical analyses of money, and its evolution, continue to focus on commodity forms of money (Smithin, 2004, 2–5; Orrell and Chlupaty, 2016, 26). In these analyses, money is construed simply as a means of exchange. Moreover, emerging in the marketplace, money is thought to be neutral, as it is simply "a veil" masking exchange and has no agency of its own (Ingham, 2004, 45–46; Smithin, 2004, 6–7).

7.4.2 The Critics

Such an account is not only erroneous from a historical perspective; it also fails to provide a basis for analyzing wide-ranging monetary systems and their impacts (Smithin, 2004, 19).

More evidence-based, archaeological, and historically oriented scholars have pointed out several deficiencies in its logic. Randall Wray's (1999) critique is especially telling in this regard. Noting that the choice of a medium of exchange requires common consent, Wray questions whether, and how, early societies might have coordinated the building of a consensus, favoring one coin over another, especially given the lack of a preexisting market. How, he asks, could an agreement have come about, given the broad range of candidate objects ranging from barley, cattle, porpoise teeth, sea shells, to various metals, each of which might have served—and often did—as a monetary standard (Wray, 1999; Peacock, 2013, 31).

Beginning with the Chartalists, monetary scholars have contended that, in contrast to the work of an invisible hand, coinage was established through a top-down process, via the imprimatur of some legitimate authority (Peacock, 2013). Drawing on a broad spectrum of disciplines, these scholars have shown that money and coinage are far more multifaceted than neoclassical economists have contended (Davis, 2010). As economic geographers Ash Amin and Nigel Thrift tell us, the origin of coins is rooted not in the marketplace but rather in social and cultural relationships (Amin and Thrift, 1996; see also Duby, 1998; Simmel, 2004; Grewal, 2008; Davis, 2010, 24; Peacock, 2013; Martin, 2015, 27). Thus, in many cases, money emerged as a means of governing societal interactions and cultural practices long before it was used in market exchange. Coins were employed, for example, as bride and blood money, as ceremonial objects, and for religious purposes (Davis, 2010, 24). Moreover, the advent of coinage was far from spontaneous; rather, it was associated with the rise of institutional structures that could legitimate coins and attest to their value (Amin and Thrift, 1996; Duby, 1998; Davis, 2010, 91; Hodgson, 2015, 10).

It is clear, then, that the relationship between coinage and social structure is far more complex than previously surmised. Whereas neoclassical economists posit that coins emerged in the context of the market, historians and numismatists claim that it was standardized money that provided the infrastructure—that is to say, the platform—upon which markets evolved. Hence, market development was hardly spontaneous. Indeed, from the earliest of times, political and religious elites were instrumental both in the establishment of markets and in their regulation (Duby, 1998; Smithin, 2004, 8, 19).

To ignore the multidimensional properties of money is to bypass many of the most important and perturbing monetary questions (Dodd, 2014, 25). To address these questions, some scholar have conceived of money more holistically, viewing it as it has coevolved in conjunction with changing socio-cultural conditions (Weatherford, 1997; Duby, 1998; Davis, 2010; Peacock, 2013; Martin, 2015). Such an approach overcomes many of the problems associated with mainstream economics and, in particular, the effort to apply universal theories to site-specific problems (Hodgson, 2001; Ingham, 2004; Grewal, 2008; Davis, 2010). In contrast, looking from a comparative, evolutionary perspective at how money has changed its form and function reveals not only what is common to all money but also how monetary standards, and the different ways in which they have been created, administered, and diffused, might account for diverse impacts of considerable significance.

7.5 Monetary Standards of Value

Pursuing a socio-cultural approach, I conceive of money as a socially constructed technology. I argue that, when viewed in the abstract, money functions as a standard of value that not only links entities in a monetary system but also serves to organize how we live and interact with one another in all arenas of life (Ingham, 2004; Smithin, 2004, 25; Dodd, 2014; Hodgson, 2015; Martin, 2015, 33).

7.5.1   *Value Is Negotiated through Social Interaction*

Following Ingham and others, I view monetary standards from a sociological perspective—that is, as being socially constructed. Expounding on what such a sociological perspective might entail, Ingham (2014) claims:

...by a 'sociology of money' I intend more than the self-evident assertion that money is produced socially, is accepted by convention, is underpinned by trust, has definite social and cultural consequences and so on. Rather, I shall argue that *money is itself a social relation*; that is to say, money is a 'claim' or 'credit' that is constituted by social relations that exist independently of the production and exchange of commodities. Regardless of any form it might take, money is essentially a provisional 'promise' to pay, whose 'moneyness', as an 'institutional fact', is assigned by a description conferred by an abstract money of account. (20–21, Emphasis mine)

Money, then, is *relational*. (Emphasis mine). It entails two or more parties to an interaction or exchange (Smithin, 2004). As Dodd attests, "money's value, indeed its very existence, rests on *social relations between its users"* (Dodd, 2014, 29). Citing Simmel (2004, 68), Dodd (2014) notes:

Just as society is made possible by the act of our being social, so is value. Value is not a property that 'belongs' to an object like color, taste, or smell. It is a third category, 'which stands, so to speak, between us and the 'objects.' Although value and exchange are intersubjective processes, they operate in such a way that value comes to appear as an objective property of things themselves. (49)

Accordingly, I contend that money is best defined in the abstract as a standard interface that determines value as it emerges from within a platform of interactions. As such, it governs the way, and how, monetary interactions can take place—that is, who can be involved, under what circumstances, and according to what rules. Of course, what constitutes value differs from culture to culture and from situation to situation (Martin, 2015, 52–60). But, as Simmel has claimed, when viewed in the abstract, value can be defined as the standard forfeiture necessary to achieve some benefit (Simmel, 2004, 82–90).

7.5.2   *The Functions of Monetary Standards*

When defined as a standard of value, money can also be differentiated based on its functions (Davis, 2010, 52; Peacock, 2013; Orrell and Chlupaty, 2016). Thus, as in the case of gold bullion, which is virtually indestructible, money can serve as a store of value. Money can also be viewed as the value of one-sided settlements, such as in the payment of taxes, restitution, war money, or other such obligations. At the same time, monetary value can be ephemeral, serving as a unit of account that documents the relationship between debits and credits. And, of course, money can be construed as a medium (value) of exchange.

7.5.3   *Money's Societal Functions*

Equally important—although often disregarded—money functions as a standard interface that serves to negotiate the boundaries between diverse players and objects (Holland, 2014).

Embodying norms, laws, symbols, codes, algorithms, etc., it provides the rules, or protocols, to be followed so that entities can interact. Thus, money not only signals critical social, political, and economic information; it also determines how interactions take place, the conditions under which they occur, and their appropriateness (Dodd, 2014, 456–457).

When viewed as a social and cultural phenomenon, it is clear why money is contagious. As Orrell and Chlupaty have observed, "Money … has within it a tendency towards unification, homogenization, and the development of a single currency (such as gold)" (Orrell and Chlupaty, 2016, 43). As importantly, money tends to integrate and unify the players and activities associated with the use of money. Thus, we find that new forms of money and monetary practices emerge by borrowing and incorporating monetary standards that were made by earlier or neighboring cultures. Moreover, as geographic centers coevolved together with the circulation and adoption of innovative monetary conventions and practices, similar social transformations took place, leading in the long run to a capitalist convergence (Weatherford, 1997; Duby, 1998; Hirschman, 2013; Hodgson, 2015).

7.5.4   What Makes a Standards Approach Unique?

A standards-based perspective of money is, therefore, decidedly different from mainstream economics. Whereas the unit of analysis in mainstream economics is a "representative" rational economic agent, operating independently from all others, *from a standards perspective the unit of analysis is, in fact, the interactions between and among diverse agents* (Emphasis mine). No wonder, then, that for mainstream economists, money is considered to be a mere, neutral device, important only insofar as it facilitates exchange. In contrast, when defined as a standard of value that serves to link and mediate interactions, money is far from neutral. *As with all standards, money generates costs and benefits, and distributes them unevenly, giving rise to winners and losers.*

7.6   Money in Context

We can see, then, that monetary standards are the product of the social, cultural, political, economic, and technological contexts in which money is embedded. Over the long-term, monetary standards co-evolve together with the evolving context in which they are rooted. A key, determining feature is the structure of the societal order. Thus, to understand the rise of money over time, as well as the diverse roles it has played, we need to view money in relationship to its unique historical circumstances. Below, I described the evolution of money in five diverse contexts.

7.6.1   Homeric Greece—Where's the Money?

Money, as a standard of value, existed long before capitalism. In undifferentiated pre-market societies, social norms and the rituals that reinforced them governed the accumulation and distribution not only of material goods but also of social status and the privileges and obligations associated with them. These practices were primarily intended to organize and reinforce social, economic, and cultural relations.

Consider, for example, the world of Greece in its Dark Ages, as described in Homer's epic tales, the *Iliad* and the *Odyssey* (Weatherford, 1997, 45; Peacock, 2013; Martin, 2015, 72–85). Comprised of isolated tribal communities, the region was characterized by brutal wars and the heroes that conducted them. Absent was any role for commodity money (Weatherford, 1997, 45–46).

Quite the contrary, resources—both material and socio-cultural—were accumulated via pitted battles, according to the principle of "might is right." The war booty was distributed in keeping with ritual gift exchanges among allied tribal chieftains (Peacock, 2013, 80). On returning home from war, these chieftains would sacrifice an ox to the gods in gratitude for their victories in battle. The roasted meat would then be distributed equally among male members of the tribe. This act signaled that all male tribal members were of equal worth and subject to the same obligations. In this way, the system of material distribution was not neutral. Rather, it reinforced confidence in the prevailing social structure, as well as the long-term cohesion of the tribe (Martin, 2015, 34–37).

7.6.2   Standards of Account in Tributary Systems

As societies grew and became more complex, it was necessary to develop new means of accumulating and allocating critical resources. The first large-scale economies were tributary systems that covered vast territories and hosted diverse populations. Keeping track of the flow of tribute from the populace to the state and back again proved a daunting task, requiring innovative techniques and organizational structures. Most important were the state bureaucracies administered by government and religious authorities whose legitimacy and accounting skills enabled them to accumulate and distribute vast resources based on a system of units of account—that is to say, records of system-wide debts and assets (Ingham, 2004, 15; Peacock, 2013, 55).

Mesopotamia provides a case in point (Ingham, 2004, 114; Martin, 2015, 38–40). A wealthy riparian region, Mesopotamia was oriented around the floodplain of the Tigris and Euphrates rivers and the city of Ur. Power was situated in the palace under the authority of a semi-divine king, who served as both military leader and chief justice within the temple's religious bureaucracy. The temple's administrators were excellent record keepers, using standardized clay tablet cuneiforms to document commercial transactions (Weatherford, 1997, 52–53; Martin, 2015; Orrell and Chlupaty, 2016, 28). Martin, 2015 Thus, there was no need for money to be exchanged. In fact, the means of production were all managed and controlled by the temples and palaces (Ingham, 2004, 114).

To account for all tribute going to and from the state bureaucracy, measures of quantity and quality were established according to a system of standardized units of account—the shekel and multiples thereof. All prices, whether for commodities or restitution, were based on this standardized unit. And it was the money of account that served as the means by which society was delineated, organized, and managed (Ingham, 2004, 115–116). Hence the unit of account defined the social structure that characterized the system and determined the allocation of resources among the temples, palaces, and farmers. This system of accounting worked so well in large-scale tributary systems that it took 2000 years for coinage to replace it.

7.6.3   Commerce and Coinage: The City States

Coinage did not first appear in large-scale agricultural systems, but rather in small kingdoms and city-states (Ingham, 2004, 120). For example, coins were first produced and used in exchange in mid-6th century Lydia—the capital of Sardis—where an extraordinarily rich vein of silver ore was discovered. Galvanizing the use of these coins was the establishment of free and open retail markets where artisans and merchants prospered by selling specialized goods (Weatherford, 1997, 48–49). Coinage, and the commerce it facilitated, helped to make Sardis very rich.

However, the production of coinage shifted to Athens with the defeat of Sardis by the Persians in 547–546 BCE.

The most prominent Athenian coins, the tetradrachm, were standardized not only by weight but also by the state images featuring the head of Athena on the obverse side and the owl and olive spray on the reverse. Reliance on Athenian coinage spread together with the rise of Athenian power. The impetus was not so much commerce but rather financing warfare (Orrell and Chlupaty, 2016, 32). Because coinage enhanced the state's legitimacy and power, the Athenian state played a pivotal role in promoting its own coins (Peacock, 2006; Davis, 2010; Peacock 2013, 99). By requiring all debits and receipts to be made in Athenian coinage, the state generated a critical mass of users, who then became locked into the use of Athenian coins. Financed through coinage, Greek military adventures also fostered the currency's widespread dissemination. Conquered territories and allies not only provided bullion to make coins; they were also required to pay tribute and taxes in Greek coinage (Orrell and Chlupaty, 2016, 36; Peacock, 2013, 140).

Neighboring states held Athenian coinage in good stead, not only because it was guaranteed by the State: its standardized form was also enshrined into law in the late 4th century under the Athenian Declaration on Coinage and Standards. This declaration required that all coin within the Athenian territory be re-minted and converted into Athenian coin (Kroll, 1993; Davis, 2010, 110). Not surprisingly, by 490 BCE, Athenian mints were mass-producing the tetradrachm to meet the growing demand in the Mediterranean world. In so doing, they provided a major boost to the Athenian economy. As John Kroll (1993) describes it: "Athens" silver industry effectively functioned as an "industry of money." But, even as Athenian coinage promoted commerce and trade, it also fostered political unity at home as well as tremendous political prestige and imperial expansion abroad. As significantly, the pervasive use of Greek coins revolutionized Greek society (Martin, 2015, 55–66).

7.6.4 Linking to the Spiritual in Late Iron Age Britain

In contrast to the Greeks, who valued coinage for its commercial and political value, the people of late Iron Age Britain esteemed the symbolic and religious role that early coins played in integrating their community and sustaining their authority. Britain, during this period, was characterized by turbulent border incursions from the continent and the rise of an elite group of warring individuals that, together with their loyal body of horsemen, jockeyed among themselves for power. The leaders lavishly bestowed coins and gifts, such as rings and torcs, on their followers to recruit and retain them. Power and influence were thereupon accorded to those who exhibited the greatest wealth and number of clientele (Creighton, 2000).

Horses played a central role in the social structure, as attested to not only by a growing equestrian material culture but also by the imprinting of the man/horse's image, albeit incrementally modified over time, on the series of coins associated with this period. According to Creighton, this image legitimated the leader by metaphorically linking him to the spiritual world. Evidence suggests that, given the artistic and technical skill required to mint these coins, their production was carried out through spiritual rituals, practiced over generations, and executed by a special class of shamans. As Creighton (2000) contends:

> I believe the horse/man image denotes the right to rule through the alliance of a leader and nature, represented by the horse. Since in many ways this is a mystical union, the development of this imagery along lines associated with altered states of consciousness should not be particularly surprising. (53)

7.6.5   Coinage and the Fall of the Roman Empire

The history of money in the Roman Empire illustrates how money affected Rome's social structure, and with it the fate of the Empire. The Romans inherited their appreciation and usage of coins from the Greeks. Like their predecessors, they viewed their own coinage as a critical accessory to state power and strategically pushed for its expansion and adoption. In fact, as Weatherford notes, in contrast to tributary systems, which enjoined government officials to distribute material resources to the populace, Rome consciously employed money—*by virtue of its ability to structure relationships*—(My emphasis) to organize and direct political affairs throughout its empire (Weatherford, 1997, 66). Functioning, for example, as a system of "pay to play," the Roman Senate allocated political roles in accordance with a member's wealth and status (Heather, 2006, 146; Faulkner, 2013, 178–179).

Money served not only to configure Rome's internal affairs but also to define the Empire's geographic and political boundaries. Based primarily on landholding and agriculture and governed by a political culture steeped in patronage, the Roman Empire was not a market economy (Goldsmith, 1987; Weatherford, 1997, 67). In fact, for much of its history, trade and profit-making were disparaged (Lopez, 1976, 8–9; Coffee, 2017). Hence, coinage served not as a medium of exchange but rather as a platform for structuring interactions throughout the Empire (Weatherford, 1997, 65–66). As Orrell and Chlupaty (2016) describe:

> The army conquered foreign lands; slaves were put to work in mines extracting precious metals; millions of shiny coins were stamped out by hand every year and distributed to the army as payment (as usual, this was the government's largest expense); and conquered populations were subjected to taxes, payable in the same coins, ensured their circulation. (34–35)

Roman coins were used, moreover, for propaganda purposes, to project the unity of the Empire based on the emperor's persona (Davis, 2010, 90). Julius Caesar, for one, replaced ancestral heads with his own image. Brutus followed suit but went even further. Not only did he put his own profile on the reverse; on the obverse, he put the grisly attack on Caesar that led to Brutus's rise to power (Davis, 2010, 90).

As the Empire expanded and became more complex, the use and distribution of coinage was reminiscent of tributary systems, where resources were allocated directly from above in accordance with the needs of the Empire (Lopez, 1976, 7). Hence, notwithstanding Rome's fine gold coinage, monetization was relatively low (Ingham, 2004, 130). Significantly, unlike earlier monetary systems, which served to reinforce social and political relationships, Rome's prioritization of war and the use of money to reward its warrior class, as well as to buy off barbarians on its borders, was divisive (Heather, 2006, 19). With the extension of the Empire, the cost of warfare and the Empire's defenses greatly increased, while the rewards to be obtained from plunder experienced diminishing returns (Weatherford, 1997, 68).

The Roman redistribution of wealth undermined the stability of the regime. In the course of Rome's campaigns, a new class of men—the equestrians—emerged to carry out tax collection and the logistical aspects of warfare. Well rewarded for their efforts, they gained enough wealth to successfully compete with aristocrats, replacing them not only in the Senate but also in the highest rungs of government (Heather, 2006; Faulkner, 2013). The senators were not the only losers. Returning home from distant wars, the peasants found themselves uprooted, because the equestrians had not only appropriated their lands; they had also replaced the peasants with slaves won in battle (Lopez, 1976, 11). Not surprisingly, towards the end of the Roman Empire, money became increasingly irrelevant. As Weatherford (1997) notes:

> In the last centuries of the Roman Empire, the emperors operated without a workable currency; like the ancient empires that had preceded it, Rome turned to conscription and forced labor to meet its needs. The government often would not allow its citizens to pay taxes in the debased money that it still issued; instead, officials demanded payment in goods, crops, or labor. (75)

Historians have proffered numerous explanations for the collapse of Rome (Cantor, 1994; Heather, 2006; Homer-Dixen, 2006; Baker, 2007; Wickham, 2007, 223; Riddle, 2008, 49–52; Faulkner, 2013). The failure of Roman monetary policy was clearly among them. As described above, the pattern of expenditure and allocation of resources not only undermined the Empire's traditional social and political structure. As importantly, by expending wealth and land to purchase the loyalty and support of the encroaching barbarians, the policy created dissension among the Roman population and shifted resources and the balance of power to the barbarians. When the well of Roman money dried up, the barbarians set out, on their own, to achieve wealth and power by force and plunder (Duby, 1998).

7.7 The Monetary Infrastructure—Braudel's Thesis

In contrast to the neoclassical economic point of view, I contend—along with the Chartalists—that money does not exist in a vacuum. To the contrary, money emerges from within a scaffolding of social structures, which is a prerequisite for money's instantiation. It is the societal standards making up this scaffolding that link and define the monetary interactions of networked participants. As noted above, when societies became more complex, diverse groups and entities were required to translate and resolve differences of value and trading procedures. Hence, with the expansion of markets and the rise of a competitive state system, the role of middlemen—whether serving as bureaucrats, soldiers, traders, or bankers—became increasingly critical, not only in determining the form and functioning of the economy but also—and as importantly—in impacting the structure of society as a whole. Hence, we can say that *the infrastructure not only affects the form that money takes; it also has significant implications for the social order itself* (my emphasis).

Fernand Braudel derived a similar conclusion from his monumental historical analyses of the European economy between the 15th and 18th centuries[1] (Braudel, 1997). Based on his studies, Braudel claimed that it is the social apparatus girding economic interactions that accounts for the discrepancies between the auspicious claims of prosperity made by free market advocates and the ever-widening gap between the rich and the poor (Braudel, 1997).

What led Braudel to this conclusion? As he describes in his book, *Afterthoughts on Material Civilization and Capitalism* (1997), there is a major difference between early market economies in which participants interacted face-to-face and on equal terms, and later "capitalist" markets, in which participants were required to negotiate not only at a distance but also relatively blindly, *via* a complex social superstructure, which essentially controls and manipulates the entire market economy (Braudel, 1997, 40).

As Braudel asserts, as the economy became more complex, it assumed an increasingly hierarchical structure, such that key decisions were made in the dark, far from scrutiny, by a limited and select group of cohorts who, taking advantage of the largesse and variety of the resources afforded to them, restructured the social structure and the standards that instantiated it in their favor. Under such circumstances, the opportunities for collusion and corruption were rife. According to Braudel, it is by virtue of these skewed standards, which are embraced and furthered by the political elite, that the wide gap between the rich and the poor emerges (Braudel, 2019).

That being said, we might ask whether or not such advantages are sustainable in the long run. What, if anything, might lead to their undoing? If Braudel were alive today, I suspect his answer might be strikingly similar to those of complexity theorists. Both might agree, for example, that even though hierarchies abound, whether in the form of fitness peaks, corporations, or bureaucracies, given their hierarchical rigidities and barriers to adaptation, they would eventually collapse and/or be taken over when confronted with radical external changes provoking a phase transition. For, when existing hierarchies are undermined by changing circumstances, new fitness parameters, in the form of new standards, will emerge, generating new players, new skills, and new hierarchies, which are more capable of meeting evolving, system-wide needs.

7.8   The Robber Barons

The case of the Robber Barons illustrates the far-reaching impacts of changing standards in the United States. During the late 19th and early 20th centuries, a new class of entrepreneurs, unencumbered by traditional commercial standards, rose to the heights in the rapidly changing and relatively lawless post-Civil War era. Much as Braudel has reasoned, the Robber Barons' actions redefined the institutional framework in their favor, such that they became the one percent!

Carpe diem! Such was the mantra that activated the Robber Barons, as they inserted themselves into the United States economy in the aftermath of the Civil War (Axelrod, 2017). A far cry from traditional entrepreneurs, these assorted newcomers were mainly upstarts of meager beginnings who, notwithstanding their lack of formal education and experience, were very savvy about the structure and workings of a relatively amorphous, pre-capitalist economy. Relying on their intuition and driven by their own needs and interests, these young entrepreneurs successfully improvised along the way. In the process, they actively "made the market," rather than passively reacting to it (Moody, 1919; Josephson, 1934; Renehan, 2005; Chernow, 2010; Geisst, 2012; Axelrod, 2017). The backgrounds and personalities of these entrepreneurs were well suited to such unprecedented times. Self-serving and driven by ambition, they sought their fortunes at a time when the northern economy was booming and investment prospects were exceptional. As Foner (2015) describes:

> The war inspired economic boom slackened only slightly with the coming of the peace, and manufacturing output quickly resumed its relentless upward course. By 1873, the nation's industrial production stood seventy-five percent above its 1865 level. … In the same eight years, three million immigrants entered the country, nearly all destined for the North and West, their labor fueling the rapid growth of metropolitan centers like New York and smaller industrial cities from Paterson to Milwaukee. (187)

Growth, however, was spotty. What was needed under the circumstances was the capital necessary to finance large-scale industrial undertakings, as well as the administrative means to coordinate these activities. The future Robber Barons emerged to fill this gap.

Recognizing the need for a unified, national market, these early capitalists focused their initial efforts and investments on infrastructure projects, especially turnpikes, railroads, and shipping, in addition to other industries—such as petroleum, coal, and steel—that were essential to their industrial operations. To cover the high fixed costs associated with these industries, these upstart entrepreneurs not only had to generate massive amounts of capital; they also had to operate at very large scales. To create these conditions, financiers were needed to put together coalitions of investors as well as the administrative machinery to govern them. To meet this need, many financiers became intimately engaged in the industries they financed, especially in the case of the railroads, a fact that, by the turn of the century, aroused the alarm of congressional leaders.

Banks were the central players raising capital to underwrite the prodigious corporate loans necessary for companies to build out their businesses before returns set in. They were essentially retailers, collecting deposits and making loans based on them. Absent government reserve requirements, banks could lend depositors' money and garner interest on it at their own discretion while independently increasing the money supply and the liquidity of the economy. Increased liquidity was certainly essential for the growing economy, but it came with the considerable risk of overcapitalization and bank failures. As Moody (2019) describes:

> By the early eighties about twice as many railroad lines had been built as the country could profitably employ, and there had been issued about four times the amount of the securities that the country could pay dividends on. (24)

In their role of stockbrokers, banks and other investment houses also generated major capital outlays. In doing so, they not only brokered the relationship between investors and companies; they also put together syndicates and packaged deals and lent money to investors at considerable interest so they could make stock purchases. In effect, stockbrokers were great gamblers, betting not only on near-term economic conditions but more importantly on risky and unpredictable long-term outcomes. Because the Robber Baron's large payoffs depended on privileged information, their machinations were typically executed behind the scenes and in the utmost secrecy (Moody, 1919, 2–3). Skirting the law, they avoided accountability.

Not all financiers were alike. J. P. Morgan, for example, took pride in operating according to long-established "gentleman" banking standards. Although Morgan could be very tough when needed, his general style, and that of the House of Morgan, was atypical. For the Morgans and their associates, building their banking empire not so much on deceit, but rather on the basis of reputation and trust (Chernow, 2010; Pak, 2013). Understandably, absent a formal legal infrastructure, J. P. Morgan and his son Jack became acknowledged leaders and eventual spokesmen of the post-war business community. Fostering cooperation instead of competition, they went to great lengths to negotiate collective settlements in the form of trusts among competing adversaries and to provide a united front when responding to government inquiries.

At the other extreme were men such as Jay Gould and his associates Daniel Drew and Jim Fisk. While Gould was respected for his "genius," he was hated by most. Relishing a fight, Gould's behavior was most egregious, as he sought to divide and conquer his opponents, employing whatever means at hand—bribery, intimidation, extortion, and even violence (Cashman, 1993, 50–51). A wizard at shorting his competitors' assets, Gould went so far as to try (but failed) to corner the market for gold (Renehan, 2005). Other magnates, whose behavior may have been less abhorrent, likewise employed ruthless tactics.

How did the Robber Barons get away with this? As one might expect, the answer had to do with the context and—in this case—the absence of broad-based, agreed-upon standards. Recall that the Robber Barons emerged in an environment in which laws and normative standards were basically absent and in which politicians—even those at the highest levels of government—were accustomed to taking bribes. Hence, the Robber Barons had considerable free rein, especially in the early years of the Gilded Age, when the "captains of industry" were perceived to be the nation's heroes (Lynch, 2004). According to Axelrod (2017):

> The prevailing mediocrity of politicians during this period meant that the real doers in the public sphere were the business tycoons, who used their financial resources to buy both political support and the politicians themselves—at municipal, state, and national levels, purchasing mayors, governors, and senators by the bushel basket. The result was the emergence of an oligarchic class and an unholy alliance between business and government that may be fairly described as the triumph of crony capitalism. (542)

Because of their strong political support and the prevailing ideology of "survival of the fittest," the Robber Barons ruled, much as Braudel might have described, in secret and at a distance, from the pinnacle of their industries, which constituted the superstructure of both the economy and society as a whole. Although the Robber Barons successfully brought the requisite skills and know-how to bear on the post-Civil War economy, their talent failed them when, at the turn of the century, and in the wake of numerous scandals and bank failures, the industrialists found themselves having to redefine themselves, or fail, under a new, regulatory regime. Unable to adapt, many went into decline.

Notes

1 For his classic works, see (1992) *Civilization and Capitalism, 15th–18th Century, V. 1, The Structure of Everyday Life; V. 2* The Wheels of Commerce; and V. 3, *The Perspective of this World,* Berkeley, CA: University of California Press.

References

Amin, Ash, and Nigel Thrift (1996) *Globalization, Institutions, and Regional Development in Europe,* London: Clarendon Press.

Axelrod, Alan (2017) *The Gilded Age—1876–1912, Overture to the American Century,* New York, NY: Sterling Publishing Co.

Baker, Simon (2007) *Ancient Rome; The Rise and Fall of Empire,* London: Random House.

Braudel, Fernand (1992) *The Perspective of This World,* Berkeley, CA: University of California Press.

Braudel, Fernand (1992) *The Wheels of Commerce,* Berkeley, CA: University of California Press.

Braudel, Fernand (1997) *Afterthoughts on Material Civilization and Capitalism,* Baltimore, MD: Johns Hopkins Press.

Cantor, Norman (1994) *Civilization of the Middle Ages,* New York, NY: Harper Collins.

Cashman, Sean (1993) *America in the Gilded Age: From the Death of Lincoln to the Rise of Theodore Roosevelt,* New York, NY: New York University Press.

Chernow, Ron (2010) *The House of Morgan: An American Dynasty and the Rise of Modern Finance,* New York, NY: Grove Press.

Coffee, Neil (2017) *Gift and Gain, How Money Transformed Ancient Rome,* New York, NY: Oxford University Press.

Creighton, John (2000) *Coins and Power in Late Iron Age Britain,* Cambridge: Cambridge University Press.

Darius, Banjo (2020) *A Brief History of Commercial Capitalism,* Chicago, IL: Haymarket Books.

Davis, Glyn (2010) *History of Money*, Cardiff: University of Wales Press.

Dodd, Nigel (2014) *The Social Life of Money*, Princeton NJ: Princeton University Press.

Duby, Georges (1998) *Rural Economy and Country Life in the Medieval West*, Philadelphia, PA: University of Pennsylvania Press.

Faulkner, Neil (2013) *Rome Empire of the Eagles*, New York, NY: Routledge.

Foner, Eric (2015) *Reconstruction Edition; America's Unfinished Revolution, 1867-1877*, New York, NY: Harper Perennial Modern Classics.

Galbraith, John K. (1995) *Money—Whence It Came and Where It Went*, New York, NY: Houghton Mifflin.

Geisst, Charles R. (2012) *Wall Street, a History*, New York, NY: Oxford University Press.

Goldsmith, Raymond (1987) *Premodern Financial Systems: A Comparative Historical Study*, New York, NY: Cambridge University Press.

Grewal, David Singh (2008) *Network Power: The Social Dynamics of Globalization*, New Haven, CT: Yale University Press.

Heather, Peter (2006) *The Fall of the Roman Empire, A New History of Rome and the Barbarians*, New York, NY: Oxford University Press.

Hirschman, Albert O. (2013) *The Passions and the Interests: Political Arguments for Capitalism before Its Triumph*, Princeton, NJ: Princeton University press.

Holland, John (1995) *Emergence: From Chaos to Order*, Readington, MA: Addison-Wesley.

Holland, John (2014) *Complexity, A Very Short Introduction*, New York, NY: Oxford University Press.

Homer-Dixon, Thomas (2006) *The Upside of Down, Catastrophe, Creativity, and the Renewal of Civilization*, Canada: Vintage Books.

Ingham, Geoffrey (2004) *The Nature of Money*, Cambridge: Polity Press.

Josephson, Matthew (1934) *The Robber Barons, the Great American Capitalists*, New York, NY: Harcourt.

Kroll, John H. (1993) "What about Coinage," Chapter 8 in John H. Kroll and Thomas Munk.

Lopez, Robert S. (1976) *The Commercial Revolution of the Middle Ages, 950–1350*, New York, NY: Cambridge University Press.

Martin, Felix (2015) *The Unauthorized Biography—From Coinage to Cryptocurrencies*, London: Vintage Press.

Moody, John (1919) *The Masters of Capital, A Chronical of Wall Street*, New Haven, CT: Yale University Press.

Orrell, David, and Roman Chlupaty (2016) *The Evolution of Money*, New York, NY: Columbia University Press.

Pak, Susie J. (2013) *Gentleman Bankers: The World of J.P. Morgan*, Cambridge, MA: Harvard University Press.

Peacock, Mark (2006) "The Origin of Money in Ancient Greece: The Political Economy of Coinage and Exchange," *Cambridge Journal of Economics*, 30, 637–650.

Peacock, Mark (2013) *Introducing Money*, New York, NY: Routledge.

Renehan, Edward J. (2005) *Dark Genius of Wallstreet, The Misunderstood Life of Jay Gould*, New York, NY: Basic Books.

Riddle, John (2008) *A History of the Middle Ages, 300–1500*, Plymouth: Rowman and Littlefield Publishers, Inc.

Samuelson, Paul (1973) *Economics*, 9th edition, New York, NY: McGraw Hill.

Shiller, Robert J. (2012) *Finance and the Good Society*, Princeton, NJ: Princeton University Press.

Simmel, Georg (2004) *The Philosophy of Money*, New York, NY: Routledge.

Smithin, John (2004) *Controversies in Monetary Economics*, Cheltenham: Edward Elgar.

Weatherford, Jack M. (1997) *The History of Money*, New York, NY: Three Rivers Press.

Wickham, Chris (2007) *Framing of the Middle Ages: Europe and the Mediterranean, 400–800*, New York, NY: Oxford University Press

Wray, Randall L. (1999) "An Irreverent Overview of the History of Money from the Beginning of the Beginning Through to the Present," *Journal of Post Keynesian Economic* 21, 11.

8

STANDARDS, MODULARITY, AND INNOVATION

8.1 Introduction

"The thing that hath been is that which shall be, and that which is done is that which shall be done; and there is no new thing under the sun" (Ecclesiastics 1:9, *King James Bible*). As seen in the Old Testament, during the early Hebrew period, and well before, creativity was deemed the sole prerogative of God. Because God had established the world *ex nihilo*, all subsequent creations were considered byproducts of his handiwork. Hence, innovation was disparaged, and creators' efforts were unrewarded (Montuori and Purser, 1995; Weiner, 2000; Sawyer, 2006, 13–14).

Notwithstanding the admonitions of the Book, since the mid-17th century, the world has witnessed rapidly accelerating rates of innovation (Johnson, 2010). Speaking, in 1965, about the increasing rate of change in the semiconductor industry, Gordon E. Moore, co-founder of Intel Corporation, predicted that the number of transistors in a dense integrated circuit would double approximately every two years, a prediction that has held true at least until the present. As importantly, the magnitude of what computing power can do has likewise grown exponentially (Beinhocker, 2006, 301). Ray Kurzweil (2001) has extended Moore's notion to include all technological progress. As he claimed in his 2001 essay, "The Law of Accelerating Returns":

> The history of technology shows that technological change is exponential, contrary to the commonsense 'intuitive linear' view. So we won't experience 100 years of progress in the 21st century—it will be more like 20,000 years of progress (at today's rate).
> *(http://www.kurzweilai.net/the-law-of-accelerating-returns)*

How do we account for the exponential growth of inventiveness? Can we reconcile the biblical idea that there is nothing new under the sun with this continuous surge of creativity? Might standards solve the puzzle? Might they serve to bridge the old world and the new? Might they provide the infrastructure for evolution itself? I think so. For, according to the definition of standards laid out in Chapter 1, standards are the interfaces that connect all things, even the past and the future. (Arthur, 2009; Goldstein et al., 2010; Johnson, 2010, 46; van Schewick, 2012; Hidalgo, 2015).

DOI: 10.4324/9781032721125-12

8.2   Modular Building Blocks

In his book, *The Nature of Technology: What It Is and How It Evolves* (2009*)*, Brian Arthur accounts for how technology spawns perpetual innovations. As he claims, technology begets itself. Much like complexity hierarchies, technologies are composed of a central assembly of interconnected technologies, each supported by several sub-assemblies of technologies that, in turn, are sustained by sub-sub-assemblies of technologies on down to the last module (Beinhocker, 2006, 263; Arthur, 2009, 32–35). As in lego blocks, these components have standard interfaces, so they work seamlessly with one another and can be reconfigured and used to build other, new technologies (Arthur, 2009, 42, 43). Over time, these modular units serve as mechanisms of heredity, linking the past and the present (Arthur, 2009, 55). By combining standardized modules in new ways, innovative technologies can be created that serve new purposes. Hence, the greater the store of modules, the more possibilities there are for assembling new technologies (Beinhocker, 2006, 263; Arthur, 2009, 37; Johnson, 2010). As Arthur opines, when technology is so conceived, it appears to be not just a working object, but rather a whole ecology of objects that evolve in novel ways over time through their combination and recombination (Arthur, 2009).

Arthur's insights apply not only to technologies, but also, as described below, to all other aspects of life. Much like technologies, modular life forms are sustained and upgraded via standards as they interact upon a network platform from which they emerge at a higher level of fitness.

8.3   The Role of Platforms

As standardized components evolve and interconnect, they create platforms upon which higher level, and more complex technologies can flourish and interact[1] (Arthur, 2009; Gawer, 2009; Johnson, 2010). These platforms serve as hubs for the convergence and interaction of diverse actors/actants[2] whose purposes and functions align. By providing an infrastructure that furnishes space for the agglomeration of new resources and diverse information, platforms are essential for innovation. Moreover, because they are comprised of standardized components, their resources can be reused to create new assets, thereby generating economies of scale and scope as well as positive externalities. As described by Baldwin and Woodword (2009):

> A platform architecture partitions a system into stable core components and variable peripheral components. By promoting the reuse of core components, such partitioning can reduce the cost of variety and innovation at the system level. The whole system does not have to be invented or rebuilt from scratch to generate a new product to accommodate heterogeneous tastes or respond to change in the external environment. (19)

Platforms are also complex adaptive systems, as described in Chapter 1. As in all such systems, standards define the structure, or architecture, of a platform, determining not only its boundaries—that is, who can access it—but also the modes and means by which interactions can take place within it (Holland, 2014). In the process, standards define the fitness criteria that allows successful innovations to rise up the fitness landscape to a higher level platform.

8.3.1   Consider the Internet

To appreciate how standardized platforms foster value-adding innovations, consider the Internet.[3] Its architecture, and the standards that define it, are governed by design principles calling for a modular, layered, end-to-end construction (Saltzer et al, 1984; Clark, 1988; van Schewick, 2012). It is this loosely coupled construction that accounts, to a considerable degree, for the Internet's dynamic, innovative evolution.

The early Internet was founded on a consensus-based, universal TCP/IP protocol suite, developed by the Defense Department in a bottom-up process under the auspices of the Advanced Research Project Agency (ARPA) (Garcia, 2016). Designed to be interoperable among its network participants, the ARPANET fostered resource sharing among defense contractors as well as greater productivity and innovation (Bolt and Newman, 1981, 11–12; Clark, 1988; Abbate, 2000). To enhance its diversity, ARPA was decentralized and organized in a somewhat fluid fashion, so that roles and relationships were negotiated in an ongoing practice (Hauben, 1996; Ryan, 2010; Garcia, 2016). Projects were distributed among major research centers, allowing each to deal with diverse, albeit complementary, strands of the work (Hafner and Lyon, 1996, 46; Abbate, 2000). The Network Working Group, a core group of participants dedicated to pursuing the ARPANET vision, coordinated these efforts (Abbate, 2000; Garcia, 2016).

This bottom-up, distributed process, and the interoperable suite of standards sustaining it, provided a platform upon which higher level services could be seamlessly overlaid (Clark, 1988; Baldwin and Woodword, 2009, 23). As importantly, the Internet's modular, layered architecture based on these standards, enabled, as well as induced, independent innovators to develop additional, innovative components that generated positive externalities and a critical mass of users (Garcia, 2016). Included among these applications were browsers, e-mail, file sharing, voice telephony, and streaming video, to name but a few (van Schewick, 2012; Garcia, 2016).

8.3.2 Nature's Platforms

Like the Internet, the biological world depends on platforms to enhance its innovative capacity[4] (Levin, 1998). Consider the coral atolls that Darwin encountered on the Keeling Islands during his Voyage of the Beagle. It was here that Darwin discovered how calcareous coral shells created a skeletal platform that housed and fed a multitude of ocean creatures. By building atop of the platform, these creatures were able to provide housing and resources for still other, higher level, forms of life (Johnson, 2004).

Beavers are also platform builders. Described by some as "wetland system engineers," beavers use their teeth, webbed feet, and rudder-like tails to build dams that provide accommodations for other wildlife, including fish, turtles, frogs, birds, ducks, and elk (*Beaversprite News Magazine,* nd; Johnson, 2004; Goldfarb, 2018). In so doing, they foster the emergence of wetlands. Beaver dams, for example, attenuate the flow of water, thereby raising the water table. The result is a boggy riparian ecosystem endowed with a whole new platform of diverse plant and animal species (*Beaversprite News Magazine* nd).

8.4 Earthly Innovation in a Complex Environment

Viewing our planet Earth as a multi-tiered platform, we can appreciate why standards were essential for the emergence of life. In the century and a half since Darwin first posited that life might have originated in some "small warm pond" where "a protein compound was chemically formed, ready to undergo more chemical changes …," researchers have proposed that life began in a primordial soup in which chemicals adhered to one another via a common interface as they bumped around in a sloshing ocean (Marshall, 2016). Carbon—a platform in its own right—was thought to play a major role in this process because carbon atoms have four electronic valences on their outer shell, making them extremely good at forming interconnections (Johnson, 2004, 27–28; Padgett and Powell, 2012).

Lack of direct evidence to support this proposition has led scientists to advance several competing hypotheses that, although reasoned along the same lines, differed with respect to what set off the emergent processes that generated the primary components of life (Johnson, 2004, 8–29; Padgett and Powell, 2012; Marshall, 2016). However, these positions have coexisted in a general perspective, which argues that the capacity for replication, metabolism, and energy—all necessary for life—simultaneously came together, interacting in shallow volcanic pools rich in metals, where the sun's ultraviolet rays could spark an autocatalytic, chemical reaction, thereby creating a platform for life's emergence (Kauffman, 1995, 42–61; Padgett and Powell, 2012; Marshall, 2016).

8.5   Standards and the Rise and Fall of Platforms

Standards not only provided the basis for autocatalysis, as in the case of the carbon atom. According to complexity biologist Stuart Kauffman (1995), standards platforms also allow continual incremental changes to occur without triggering constant disrupting events. As he notes, if changes in species were to occur in an unstable environment, chaos would ensue. On the other hand, a rigid environment would preclude any changes at all. As Kauffman argues, evolution and adaptation take place at the edge of randomness and order, a condition that standards platforms—given their relative stability but potential adaptability—provide (Kauffman, 1995, 40; Johnson, 2004, 30; Baldwin and Woodword, 2009, 23–24; Beggs, 2022).

As we have seen in Chapter 5, this is not to say that massive disruptions, such as avalanches, volcanic eruptions, or political revolutions, do not occur (Saviotti, 2009, 27; Scheffer, 2009). They do, in what is characterized as a phase transition (Bak, 1999; Buchanan, 2000; Beinhocker, 2006; Chapter 7). But such events are far rarer that incremental changes. In fact, they follow a power law: that is to say, according to some ratio, the larger the event the less frequent its occurrence. As importantly, it is during phase transitions, when standardized rules are upended, that ruptures in the system usher in new standards accompanied by major innovative changes (Bak, 1999)

Eric Weiner's work provides evidence supporting Kauffman's observation that creativity and innovation occur at the edge of chaos. In his book *The Geography of Genius* (2016), Weiner documents how the greatest creative breakthroughs in history have occurred in places that were experiencing fragmentation and chaos (Weiner, 2016, 31). Pointing to the situation in Athens at the time of Socrates, he notes:

> During this period, known as the Axial Age, old orders were crumbling, and new ones were not yet solidified. Cracks appeared, and as it's been said, the cracks are what let the light in.
> *(Weiner, 2016, 45–46)*

The same can be said for the period of the Renaissance, as described in Chapter 11.

8.6   Standardizing Information—A Critical Resource

Modules and modular architectures not only provide standard building blocks upon which learning and innovative emergence can take place; they also help to capture and embed new information and knowhow required for adaptation and ascent up fitness landscape. This, then, begs the question of how information might be standardized, so as to be useful to innovators.

Cesar Hidalgo (2015) offers a compelling theory. As he points out in his book, *Why Information Grows, The Evolution of Order from Atoms to Economies*, information is ephemeral. It emerges spontaneously in non-equilibrium systems moving towards a steady state. For example, if you heat water in a pan on the stove, it will begin to churn as in a non-equilibrium system. But, when it reaches a boiling point, it will take on a physical, hexagonal structure, which manifests information. However, because information is subject to the second law of thermodynamics, this structure will dissipate over time. Were the heat to be extinguished, the water would return to its randomized, fluid state (Buchanan, 2000, 80; Goldstein et al., 2010, 79–82). To preserve information and knowhow, so as to make it available as a standardized input for future innovations, it must be embodied in a physical structure, a state Hidalgo characterizes as "crystallized imagination" (Hidalgo, 2015; see also Lane, 2009, 16).

Yet not all physical objects contain a significant store of information and knowhow. For this to happen, two more conditions in addition to physicality must be met. First, the physical configuration cannot be randomly assembled. To contain a significant amount of information, it must be highly structured. Secondly, while a physical casing can protect information from the forces of entropy, matter can only build and expand upon this information if it can compute (Hodgson and Knudsen, 2013; Hidalgo, 2015, 3). How does such computation occur? Once again, we see that standards—in this case, in the form of codes—are critical to the operation. Take a seed, for instance. It not only consists of information in its genetic material; it also contains diverse organelles that allow the coded information in the DNA molecule to be accessed, unpacked, or reproduced (Hidalgo, 2015, 86–87). Similarly, a fertilized bird's egg embodies the code to convert the DNA for a chick into a living chick, as does the fertilized human ovum to reproduce a human being (Beinhocker, 2006, 260). Chemical catalysts similarly carry out computational functions, containing within them the information and machinery needed to speed up reaction times.

8.7   Structuring Outcomes *via* Network Architecture

We have seen that standardized information is embedded in standardized physical modules, which provide the building blocks for innovation. As well, we have noted how standard codes embed the information required to replicate and even enhance these building blocks. Equally important, standards determine the architecture of a network—that is, how network modules are configured and interact. Whether or not network structures are conducive to fostering innovation will depend not only on the context in which the network is situated but also on the network architecture, as defined by the standards governing it (Burt, 2007).

Recognizing the strategic value of networks in fostering innovation, scholars and business analysts have sought to identify the network architecture most likely to promote it. Their inquiries have engendered social network analysis, a growing field of research (Borgatti et al., 2009; Easley, 2010). From these analyses, two lines of reasoning have emerged: one emphasizes the importance of weak ties and brokers in the network, whereas the other focuses on the value of cohesion and strong ties. Increasingly, these two strands have been brought together around the concept of a "small world network."[5]

8.7.1   Weak Ties

The focus on weak ties dates to 1973, with the publication of Mark Granovetter's groundbreaking study, "The Strength of Weak Ties" (Granovetter, 2001). Examining the role of network connections in "getting a job", Granovetter argues that, when seeking new information—such as that about job opportunities—it is better to work through acquaintances, or weak ties, rather than through close relationships, or strong ties. As Granovetter argues, when people are tightly linked through strong reciprocal ties, their views and the information they possess are typically redundant. In contrast, weak ties to external individuals and diverse groups provide access to new information and knowledge resources (Granovetter, 2001). Pursuing this thesis, others have pointed out that, when actors are densely connected in close-knit clusters, social pressures to maintain the status quo will likely inhibit innovative ideas (Burt, 1992; Rogers, 1993; Tsai and Ghoshal, 1998; Burt, 2007; Easley and Kleinberg, 2010).

8.7.2 Network Brokers

Expanding further on the notion of weak ties, Ron Burt (1992, 2007) developed the concept of a network broker, an ideally situated weak tie that spans "structural holes"—that is, empty spaces between two or more diverse clusters. Occupying such a position, brokers not only have privileged access to diverse information and resources, but they can also capitalize on their brokerage positions to create new opportunities by channeling resources to the clusters where they can be best exploited (Burt, 1992, 2007; Fleming and Marx, 2006; Easley and Kleinberg, 2010).

Of course, simply having access to knowledge and resources does not, in and of itself, yield creative outcomes. To produce innovative results on a recognizable scale, individuals must act collectively (Sawyer, 2006; Lazer and Friedman, 2007; Easley and Kleinberg, 2010; Goldstein et al., 2010). Achieving such collective action is often problematic, however, due to the costs associated with organizing and administering collaborative behavior as well as to the problems and potential contentiousness entailed in allocating and distributing collective benefits[6] (Olsen, 1971; Cornes and Sandler, 1986; Gulati and Singh, 1998).

8.7.3 Strong Ties

To understand how network architecture might promote the collective action required to maximize the value of information and diverse skill sets, some scholars have focused on the impact of clusters and strong ties. These researchers found that, to overcome the difficulties of achieving collective action, participants in a joint endeavor must be well connected. For, when innovative initiatives are organized in densely connected clusters, innovative information, and the influence associated with it, can travel rapidly along multiple paths (Krackhardt, 1992). As importantly, because clusters are comprised of like-minded people, actors that are engaged in them tend to generate a common identity, which facilitates collective action (Coleman, 1988; Passy, 2003, Collins, 2004; Burt, 2005). Close-knit groups also provide transparency and generate norms of behavior that reduce opportunism and enhance trust, essential ingredients of creative collaboration (Gould, 1993; Uzzi, 1996; Uzzi et al., 2007).

8.7.4 Small World Networks

These two ways of depicting innovative network architectures are by no means mutually exclusive. In fact, many analysts have woven these two theoretical strands together with the concept of "small world networks". A small world network is one that exhibits both high levels of clustering together with weak links that provide a short path length across the network (Watts D.1999; Buchanan, 2002, Newman, Barabasi and Watts, 2003; Uzzi et al., 2007; Borgatti et al. 2009). Conceived as such, a small world network combines the benefits of both clusters and weak ties. As innovation and creativity scholars explain, weak ties linked by small path lengths across the network enhance search capabilities and allow new information to flow quickly throughout the network, while clusters provide absorptive capacity and the level of collaboration required to take advantage of it (Burt, 1992; Fleming and Marx, 2006; Burt, 2007; Uzzi et al., 2007; Borgatti et al., 2009).

Not surprisingly, given these benefits, small worlds can be found among any number of systems, both natural and man-made. These include electric grids, world airports, the World Wide Web, business alliances, and the worm *Caenorhabditis elegans,* to name a few (Buchanan, 2002; Newman, Barabasi and Watts, 2003, Uzzi et al., 2007). As telling, the widespread existence of small worlds appears to be an emergent phenomenon insofar as small worlds are not so much purposefully designed, but rather are, to a considerable degree, the product of self-organization and chance (Buchanan, 2002; Fleming and Marx, 2006; Uzzi et al., 2007; Goldstein et al., 2010).

An additional factor merits attention in explaining small world performance. This is the level of network clustering, a variable elaborated upon by Stuart Kauffman in his book, *At Home in the Universe* (1995). Too many connections can give rise to what Kauffman refers to as "connectivity catastrophes." Subsequent empirical research has shown that the benefits of interconnection follow a linear progression but only up to a certain threshold. Much as Kauffman (1993) has argued, the ideal network is one in which connectivity falls within a medium range (Lazer and Friedman, 2007; Uzzi et al., 2007, 87; Beggs, 2022).

8.8   Organizational Restructuring for Innovation

As one might expect, just as technologies and nature's systems have evolved to enhance their fitness in response to their changing environments, so today's organizations must transform themselves to operate successfully as their world becomes increasingly complex (Padgett and Powell, 2012; Garcia and Fischer, 2023). To do so, organizations must reorganize the ways in which their component parts interact (Garcia, 2015). To illustrate the relationships between rearranged network components and organizational outcomes, Garcia compares two natural phenomena: graphite and diamond. As he notes:

> Graphite is common, soft, and opaque. In contrast, diamond is rare, hard, and translucent. What is amazing is that both graphite and diamond are made entirely of carbon. How can two substances that are so different be comprise of the exact same thing?

> The answer is connections. What makes graphite and diamond different is not the carbon from which they are made but how the carbon atoms are connected. In graphite, carbon atoms are connected in loosely coupled sheets that easily slough off. This is why graphite is often used as a lubricant in industrial applications. In diamond, on the other hand, the carbon atoms are connected in every which way and result in some of the strongest bonds in the universe. In short, connections matter.

> *(Garcia, 2015, 10–11)*

As the analogy suggests, the structure of connections within an organization is likewise a major determinant of how it performs (Padgett and Powell, 2012; Garcia, 2015). It should come as no surprise, then, that firms are increasingly seeking to enhance their innovative potential not only by reconfiguring modular, standardized components but also by building network platforms based on specialized, modular production processes that extend outward to additional resources and information, creating thereby a small world (Langlois and Robertson, 1992; OTA, 2004; Singh, 2005; Fleming and Marx, 2006; Goldstein et al., 2010; Garcia, 2015). These inter-firm network platforms, which typically entail knowledge sharing and collaborative R&D, are most prevalent in high-tech industries that operate in highly uncertain environments. In fact, as the empirical evidence confirms, these network platforms serve as the "locus of innovation." They not only resemble small worlds; they also rank high in terms of innovative performance (Zirulia, 2009).

In their book *The Emergence of Organizations and Market* (2012), Padgett and Powell (2012) build on the biological metaphor of autocatalysis[7] to explain how innovative changes are brought about by the restructuring of standardized resources and information. As they explain:

> For us, emergence of organization is grounded in transformations of social networks, which wend through organizations, bringing them to life. From autocatalysis we appropriate a commitment to discovering and formalizing processual mechanisms of genesis and catalysis, which generate self-organization in highly interactive systems .... Learning at the human level is equivalent to co-evolution of rules and protocols at the 'chemical' level. Actors thereby become vehicles through which autocatalytic life self-organizes.
>
> *(Padgett and Powell, 2012, 27–28)*

As the authors argue, multiple networks, overlapping in a heterarchical fashion (Kontopoulos, 1993), provide a context for each other, allowing organizations to incorporate components from previous types of organizational entities. These are then passed on via standardized production rules and protocols of interaction. The resultant sharing of knowledge and practices across networks generates new standards of behavior that, over time, form the basis of innovative business practices and culture. (Padgett and Powell, 2012, 28–36). The transference of standardized skills and their embodiment and reproduction in people is, according to the authors, essential to this process (Padgett and Powell, 2012, 140).

8.9 Social Transformations

Large scale social transformations similarly occur in response to the evolution of social structures (Bak, 1999 12–13; Sole, 2012). How do these transformations come about? As complexity theory purports, the social order is made up of hierarchical layers, such that each successive layer upward on the fitness landscape becomes increasingly complex as it incorporates greater information and resources.[8] Managing such complexity requires greater control, leading to bureaucratization and the tightening of, and increased rigidity among, interrelationships. As in the case of technologies, we might call this situation "lock-in". Societal lock-in leads to a "critical state,"[9] at the edge of chaos, which is subject to collapse when radically new circumstances emerge, and the system is unable to successfully adapt to them[10] (Scheffer, 2009). As Schaffer (2009) describes:

> ...Social systems are notorious for periods of relative inertia with occasional rapid transitions on scales varying from locally held opinions and attitudes to massive shifts such as

the collapse of states and civilizations. …Just as a ship can be unstable if too much cargo is loaded on deck, complex systems ranging from the climate to ecosystems and society can lose resilience until even a minor perturbation can push them over a tipping point. While some critical transitions can play havoc on society others represent escapes from undesirable situations. (13–14)

Perhaps it is this scenario that accounts for why such collapses characteristically occur in those societies that are the most complex and operating at the top of their fitness levels.[11]

But the decline of a system is only half of the story. Phase transitions leave a residue of societal fragments that, together with new actors and resources, can be agglomerated to construct more complex, and adaptive, systems that function at a higher level of fitness (Schwartz and Nichols, 2010). For a recent, captivating example, consider how today many small Italian towns are seeking to refurbish their hollowed-out communities by offering decrepit housing properties to foreigners for minimal amounts of currency in exchange for their overhauling them. Buyers from around the world have taken up these offers, and to good effect. As a result of their efforts, these new, and diverse, homeowners have not only turned these properties into stunning abodes; they have also, in the process, reinvigorated and revamped these communities by ushering in a wide range of new ideas and resources (https://www.cnbc.com/2024/04/06/best-advice-on-buying-a-1-home-in-italy-from-people-who-did-it.html).

Padgett and Powell offer a theoretical explanation of how such societal transformations occur (Padgett and Powell, 2012; Padgett, 2016). Conceiving the world as a layered system composed of fluid networks, each of which constitutes a given domain of activity, such as the military, the economy, kinship groups, etc., they posit that societal transformations come about when actors, operating across multiple networks, borrow standardized modules of behavior from different realms, and transpose them in an autocatalytic process, so as to generate new, innovative forms of social life. Such transpositions occur when systems are vulnerable due to some external perturbation. As Padgett (2016) notes:

A shock—sometimes exogenous (like a war or plague), sometimes endogenous (like a revolt or business collapse)—is often (not always) observed to scramble or weaken not just a single node but a whole subset of notes, thereby creating temporary vulnerability in those nodes' network reproduction. The mechanics of constructive "rewirings" observed to transpire within such temporal and network windows of vulnerability, moreover, often involve transpositions of people and their network practices from one domain to another. (195)

8.10   City-Wide Hubs of Innovation

Much like the fluid networks described by Padgett and Powell (2012) cities are comprised of multiple intersecting networks that breed societal innovations in the face of disruptive changes. It should come as no surprise, then, that when cities emerged in the 13th century, they provided the motor for Europe's economic growth. As French historian Braudel claimed in his monumental three volume study, *Civilization and Capitalism, 15th-18th Century* (1992):

… it was the medieval city—a more or less active ferment depending on period and place—which, like the yeast in some mighty dough, brought about the rise of Europe …. the crucial move was made from a domestic to a market economy. In other words, the towns were beginning to tower above their rural surroundings and to look beyond their immediate horizons. This was a great 'leap forward,' the first in a series that created European Society and launched it on its successful career.

(Braudel, 1992, 94)

What made these cities so dynamic was their small world architectures. At their core, they consisted of the full range of institutions required for capitalistic exchange (van Bavel et al., 2011). At the same time, these cities benefited from their extended networks outward, which allowed them to capitalize on the greatly expanding European economy. Together, the diversity generated by expanding trade, and the pull derived from the density of interactions yielding economies of agglomeration,[12] fostered cumulative causation, so that European cities, each in their own turn, became the central locus of the European economy. Braudel described these sites as the warehouses of the world (Braudel, 1992, 125).

Venice, the first of these cities to emerge, provides a prime example. As Braudel recounts, in contrast to other cities whose trade was primarily limited to other European cities with less developed markets,[13] Venice had the advantage of having developed a rich trading network with cities in the Levant. According to Braudel, this was a conscious policy on Venice's part. As he notes:

[it] can hardly be doubted, since she forced it upon all the cities more or less dependent upon her. All trade to and from the *Terra Firma*, all exports from her island in the Levant or cities in the Adriatic (even goods travelling to Sicily or England) were obliged to pass through the port of Venice. Thus Venice had quite deliberately ensnared all the surrounding subject economies including the German economy, for her own profit: she drew her living from them, preventing them from acting freely and according to their own lights.

(Braudel, 1992, 125)

Equally important to Venice's success was her ability to generate agglomeration economies by drawing together in close proximity all of the institutional components required for exchange. The center of this activity was the Rialto Bridge, located around the corner from the *Calle della Sicurta*. As Braudel (1992) points out:

All major business matter were therefore handled literally in the streets surrounding the bridge. If a merchant was 'deprived of his right to go to the Rialto,' this punishment signified 'as numerous appeals indicate that he was deprived of the right to participate in big business.' (129)

Because the size of the European market was finite, there was room for only one city to dominate at a time (Braudel, 1992, 89–174). Given the high stakes involved, cities fiercely competed to assume this commanding role. However, as in most complex systems, the ascendant city was, at its height, at the edge of a phase transition. For, as the European economic and political situation underwent radical changes, leading cities, locked into their traditional ways of operating, went into decline, making way for an upcoming city to insert itself in the vacuum, and thereby rise to prominence. As a result, Europe witnessed a succession of dominant cities from Venice to Lisbon, to Antwerp, to Genoa, to Amsterdam, each operating at a higher level of fitness. Amsterdam—the last of these cities—was the most powerful of all, due to its interlocking networks and their extensive reach. As one might expect, it also experienced the greatest fall. As Braudel (1992) describes:

> In Amsterdam, everything was crammed together, concentrated: the ships in the harbour, wedged as tightly as herrings in a case, the lighters plying up and down the canals, the merchants who thronged to the Bourse, and the good which piled up in the warehouses only to pour over them. ... the warehouse of Amsterdam could absorb and then disgorge an amount of goods and services on the market, all available at a moment's notice. At any given command, the entire machine went into action. This was the means whereby Amsterdam maintained her superiority—an abundance of every-ready goods and a great mass of money in constant circulation. (236)

Understandably, it was Amsterdam, the most dominant and intricately linked city that collapsed from its pinnacle in the 18th century due to the economic and political rigidities associated with its success, making it unfit to address the multitude of disruptive changes taking place within its environment. This phase transition not only signaled the deterioration of Amsterdam's dominance; as significantly, it also led to the decline of the city-centered world economy. Benefiting from a more expansive and dynamic world economy, and having established its own internal, and superior, network to service it, England became the first nation state to assume the leadership role (Braudel, 1992, 353–385).

Notes

1 See Chapter 5 for a detailed characterization of platforms.
2 The word "actant" is derived from Actor Network Theory. It refers to technologies that have agency of their own. See Bruno Latour (2007).
3 The following discussion on the Internet is drawn from Garcia (2016).
4 See Chapter 6 for a characterization of the role of the platform in the evolutionary process.
5 For an additional discussion and examples of small world networks, see Chapter 10.
6 For detailed examples of the problems associated with collective action, see Chapter 2 and Chapter 7.
7 The term catalysis refers to a change and increase of a chemical reaction. It is the process of catalysis that leads to radical changes in a phenomenon, and hence the emergence of something new. Autocatalysis as employed by Pagett and Powell refers to: "a set of nodes and transformations in which all nodes are reconstructed through [feedback and] transformations among nodes in the set." (Padgett and Powell, 2012, 33).
8 Vertical complexity refers to hierarchical governance, whereby decision-making power is centralized. On the other hand, horizontal complexity entails the differentiation of a population into various roles or subgroups.
9 Critical points are the points between two physical states which are poised to flip given the right circumstances. They exist at the cusp between complexity and chaos. It is likewise referred to as a 'tipping point.' (Kauffman, 1995 Gladwell, 2002).
10 Resilience might be hindered by factors such as power pushbacks, path dependence, and/or a lack of diversity.
11 As Lozny notes: "The more ordered systems become, the more efficient in problem-solving they are, but there is a price to pay—the more complex a system becomes, the more likely it is that the value of some critical variable will radically change, causing stress and even failure" (Lozny, 2017, 37).
12 As Allan Scott describes, "... agglomeration economies in simple market systems flow from things like shared infrastructure, the pooling of market information, and the joint attractive power of multiple seller; and it is phenomena like these that ultimately underpin the formation of clusters of market activity over geographic space. If we introduce production as well as exchange into the analysis, we will

find that the incentives to agglomeration are multiplied many times over, and the scene is now set for a vastly expanded and intensified urban process (Scott, 2013, 16).

13 As Braudel notes, "The Mediterranean and the active part of Europe were reduced more than ever to an 'archipelago' of cities (Braudel, 1992, 2, 11).

References

Abbate, J. (2000) *Inventing the Internet*, Cambridge, MA: MIT Press.

Arthur, Brian (2009) *The Evolution of Technology: What It Is and How It Works*, New York, NY: Free Press.

Arthur, Brian (1983) *On Competing Technologies and Historical Small Events: The Dynamics of Choice Under Increasing Returns*, Mimeo, Stanford University.

Bak, Per (1999) *How Nature Works: The Science of Self-Organized Criticality*, New York, NY: Copernicus/ Springer.

Baldwin, Carliss Y., and C. Jason Woodword (2009) "The Architecture of Platforms: A Unified View," in Gawer, Annabelle, ed. *Platforms, Markets and Innovation*, Cheltenham: Edward Elgar.

Beaversprite News Magazine (nd). accessed 7, May 2018 at http://www.beaversww.org/beavers-and-wetlands/vital-wetlands/

Beggs, John M. (2022) *The Cortex and the Critical Point: Understanding the Power of Emergence*, Cambridge, MA: MIT Press.

Beinhocker, Eric (2006) *The Origin of Wealth, Evolution, Complexity, and the Radical Remaking of Economics*, Cambridge, MA: Harvard Business School Press.

Bolt, Beranek, and Newman, Inc (1981) *A History of the Internet: The First Decade*, Arlington, VA: DARPA.

Borgatti, S.P., A. Mehtra, D.J. Brass, and G. Labianca (2009) "Network Analysis in the Social Sciences," *Science* 323 (892), 1–5.

Braudel, Fernand (1992) *Perspective of the World, v. 3, Civilization and Capitalism, 15th to 18th Century*, Berkeley, CA: University of California Press.

Bruno Latour (2007) *Reassembling the Social: An Introduction to Actor-Network Theory*, New York, NY: Oxford University Press.

Buchanan, Mark (2000) *Ubiquity, Why Catastrophes Happen?* New York, NY: Three Rivers Press.

Buchanan, Mark (2002) *Small World Networks and the Groundbreaking Science of Networks*, New York, NY: W.W. Norton & Company.

Burt, R.S. (1992) *Structural Holes: The Social Structure of Competition*, Cambridge, MA: Harvard University Press.

Burt, R.S. (2007) *Brokerage and Closure: An Introduction to Social Capital*, New York, NY: Oxford University Press.

Clark, D. D. (1988) "The Design Philosophy of the DARPA Internet protocols," *Computer Communications Review* 18 (4), 106–114.

Coleman, J.S. (1988) "Social Capital and the Creation of Human Capital," *American Journal of Sociology* 94(Supplement), S95–S120.

Collins, Randall (2004) *Interactional Ritual Chains*, Princeton, NJ: Princeton University Press.

Cornes and Sandler (1986) *The Theory of Externalities, Public Goods and Club Goods*, New York, NY: Cambridge University Press.

Easley, David, and Jon Kleinbeg (2010) *Networks, Crowds, and Markets: Reasoning About a Highly Connected World*, New York, NY: Cambridge University Press.

Ecclesiastics 1:9, in *King James Bible*.

Fleming, Lee, and Matt Marx (2006) "Managing Creativity in Small Worlds," *California Management Review* 48 (4), 1–22.

Garcia, Dorothy Linda (2016) "The Evolution of the Internet—A Socioeconomic Account," in Bauer, J. M., and M. Latzer, eds. (2016) *Handbook on the Economics of the Internet*, Cheltenham, UK and Northampton, MA: Edward Elgar.

Garcia, Stephen (2015) "Improving Innovation with Organizational Network Analysis," *OD Practitioner* 47 (2), 10–16.

Garcia, Steve, and Dan Fischer (2023) *The End of Leadership as We Know It: What It Takes To Lead I Today's Volatile and Complex World*, Hoboken, NJ: Wiley.

Gawer, Annabelle, ed. (2009) *Platforms, Markets and Innovation*, Cheltenham: Edward Elgar.

Gladwell, Malcolm (2002) *The Tipping Point: How Little Things Can Make a Big Difference*, Boston, MA: Little Brown and Company.

Goldfarb, Ben (2018) *Eager: The Surprising Secret Life of Beavers and Why They Matter*, Chelsea, VA: Chelsea Green Publishing.

Goldstein, J. J. Hazy, and B. Lichtenstein (2010) *Complexity and the Nexus of Leadership, Leveraging Nonlinear Science to Create Ecologies of Innovation*, New York, NY: Palgrave Macmillan.

Gould, Roger V. (1993) "Collective Action and Network Structure," *American Sociological Review* 58 (2), 182–196.

Granovetter, M., and R. Swedberg (2001) *The Sociology of Economic Life*, Boulder, CO: Westview Press.

Gulati, R., and H. Singh (1998) "The Architecture of Cooperation: Managing Coordination Costs and Appropriation Concern in Strategic Alliances," *Administrative Sciences Quarterly* 43 (4), 792.

Hafner, K., and M. Lyon (1996) *Where Wizards Stay Up Late: The Origins of the Internet*, New York, NY: Simon & Schuster.

Hauben, M. (1996), *Behind the Net: The Untold History of the Internet*, accessed 9 February 2015 at http://www.columbia.edu/~hauben/book-pdf/CHAPTER%207.pdf

Hidalgo, Cesar (2015) *Why Information Grows, The Evolution of Order from Atoms to Economies*, New York, NY: Basic Books.

Hodgson, G. M., and T. Knudsen (2013) *Darwin's Conjecture: The Search for General Principles of Social and Economic Evolution*, Chicago, IL: Chicago University Press.

Holland, John (2014) *Signals and Boundaries: Building Blocks for Complex Adaptive System*, Cambridge, MA: MIT Press.

Johnson, Steven (2004) *Emergence, The Connected Lives of Ants, Brains, Cities, and Software*, New York, NY: Scribner.

Johnson, Steven (2010) *Where Do Good Ideas Come From: The Natural History of Innovation*, New York, NY: Bloomsberg.

Kauffman, Stuart (1993) *The Origins of Order*, New York, NY: Oxford University Press.

Kauffman, Stuart (1995) *At Home in the Universe: The Search for Laws of Self-Organization and Complexity*, New York, NY: Oxford University Press.

Kontopoulos, K. M. (1993) *The Logic of Social Structure*, New York, NY: Cambridge University Press.

Krackhardt, D. (1992) "The Strength of Strong Ties: The Importance of Philos in Organizations," in Nitan Nohria, and Robert Eccles, eds. *Networks and Organizations: Structure, Forms, and Actions*, Boston, MA: Harvard Business School Press.

Kurzweil, Ray (2001) accessed 7 Feb. 2016 at http://www.kurzweilai.nt/the-law-of-accelerating-returns

Lane, David, D. Pumain, S. E. Leeuw, and G. West, eds. (2009) *Complexity Perspectives in Innovation and Social Change*, New York, NY: Springer.

Langlois, R. N., and P. L. Robertson (1992) "Networks and Innovation," *Research Policy* 21, 297–313.

Lazer, D., and A. Friedman (2007) "Network Structure of Exploration and Exploitation," *Administrative Science Quarterly* 52, 667–694.

Lozny, Ludomir R. (2017) "Societal Dynamics of Prestate Societies of the North Central European Plains, 500-1000 CE: A Model," in Richard J. Chacon and Rubn G. Mendoza, eds. *Feast, Famine, or Fighting, Studies in Human Ecology and Adaptation*, Cham: Springer International Publishing AG.

Marshall, Michael (Oct. 31, 2016) "The Secret of how life on Earth Began," BBC, http://www.bbc.com/earth/story/20161026-the-secret-of-how-life-on-earth-began

Montuori, Alfonso, and Ronald E. Purser (1995) "Deconstructing the Lone Genius Myth: Towards a Contextual View of Creativity," *Journal of Humanistic Psychology*, 35–69.

Newman, M., A.-L. Barabasi, and D. J. Watts (2011) *The Structure and Dynamics of Networks*, Princeton, NJ: Princeton University Press.

Office of Technology Assessment, U.S. Congress (1994) *Network Enterprises: Looking to the Future*, Washington, DC: Government Printing Office.

Olsen, Mancur (1971) *The Logic of Collective Action: Public Goods and the Theory of Groups*, Cambridge, MA: Harvard University Press.

Padgett, John F. (2016) "Evolvability of Organizations and Institutions," in Wilson, David S., and Alan Kirman, eds. *Complexity and Evolution: Towards a New Synthesis for Economics*, Cambridge, MA: MIT Press.

Padgett, John F., and Walter W. Powell (2012) *The Emergence of Organizations and Markets*, Princeton, NJ: Princeton University Press.

Passy, F. (2003) "Social Networks Matter. But How?" in Mario Diani, and Doug McAdam, eds. *Social Movements and Networks Relational Approaches to Collective Action*, New York, NY: Oxford University Press.

Rogers, E. (1993) *The Diffusion of Innovations*, New York, NY: The Free Press.

Ryan, J. (2010) *A History of the Internet and the Digital Future*, London: Reaktion Books.

Saltzer, Jerome, David Reed, and David Clark (1984) "End-To-End Arguments in System Design," *ACM Transactions on Computer Systems* 2(1), 277–288.

Saviotti, Pier Paolo (2009) "Networks, National Innovation Systems, and Self Organization," in Fischer and Frohlich, eds. *Knowledge, Complexity and Innovation Systems*, New York, NY: Springer, pp. 21–58.

Sawyer, Keith (2006) *Exploring Creativity: The Science of Human Innovation*, Oxford: Oxford University Press.

Scheffer, Marten (2009) *Critical Transitions in Nature and Society*, Princeton, NJ: Princeton University Press.

Schwartz, Glenn M. and Nichols, John H. (2010) *After Collapse: The Regeneration of Complex Systems*, Tucson, AZ: The University of Arizona Press.

Scott, Allen J. (2013) *A World in Emergence: City Regions in the 21st Century*, Northampton, MA: Edward Elgar.

Singh, J. (2005) "Collaborative Networks as Determinants of Knowledge Diffusion Patterns," *Management Science* 5 (15), 756–770.

Sole, Richard (2012) *Phase Transitions*, Princeton, NJ: Princeton University Press.

Tsai, W., and S. Ghoshal (1998) "Social Capital and Value Creation: The Role of Networks," *Academy of Management Journal* 41 (4), 464–476.

Uzzi, B. (1996) "Social Structure and Competition in Interfirm Networks: The Paradox of embeddedness," *American Sociological Review* 61 (4), 674–680.

Uzzi, B. F. Amaral, and Reed-Tsochas (2007) "Small World Networks and Management Science Research; A Review," *European Management Review* 4 (2), 74–91.

van Bavel, Bas, Jessica Dijkman, Erika Kuijpers, and Zuijderduijn (2011) *The Organization of Markets as a Key Case for an Institutional Approach*, CGEH Working Paper # 6.

van Schewick, Barbara (2012) *Internet Architecture and Innovation*, Cambridge, MA: MIT Press.

Watts, D. (1999) 'Networks, Dynamics, and the Small World Phenomenon,' *American Journal of Sociology*, 105 (2), 493–527. http://www.cgeh.nl/working-pper-series

Weiner, Eric (2016) *The Geography of Genius: Lessons from the World's Most Famous Places*, New York, NY: Simon and Schuster.

Weiner, Robert Paul (2000) *Creativity and Beyond: Cultures, Values and Change*, Albany, NY: State University of Stonybrook Press.

Zirulia, Lorenzo (2009) "The Dynamics of Networks and the Evolution of Industries," in Malerba, Franco and Vonortas, Nicolas, eds. *Innovative Networks in Industries*, Cheltenham: Edgar Elgar.

9

HOW STANDARDS ENGENDER TRUST

9.1   Introduction

Donald Trump, on becoming President, claimed he would repeal Obamacare, force Mexicans to pay for building a border wall, bring jobs back to the United States, establish manufacturing jobs, lower taxes for the middle classes, and establish a Muslim registry. In his words, he would "Make America Great Again." When asked by skeptics how he envisioned achieving these goals, Trump asserted: "*Trust me, I have a plan.*" And, notwithstanding numerous questions regarding his veracity and many "alternative facts," a majority of the American people have reelected him to serve a second term. How can this be? Does this mean they trust him?

Trust, you might recall, was also an issue in Hillary Clinton's 2020 presidential campaign, but in her case, it was distrust that undermined her chances. Consider that in September 2016, Clinton and Trump were tied in terms of voter trust. However, by the first week in November, Donald Trump's ranking on trust issues increased to 46 percent, eight points higher than Clinton's. This was surprising, in light of *PolitiFact*'s report claiming Trump's statements were true, or partially true, a mere 15 percent of the time. In contrast, Hillary's claims were considered 50 percent true, or mostly true (Chan, 2016). Of course, much of the distrust of Hillary was fueled by Trump himself, along with Russian hackers, whose attacks on her trustworthiness, as in "Crooked Hillary," resonated with a sizeable portion of the electorate, who saw her as not being forthright either about her email server or her perceived conflicts of interest (Gearan, 2016; Nate Silver, 2018).

The extent to which questions of trust, rather than policy, dominated the US electoral campaign in 2020 was striking (Durhally, 2016; Roberts, 2016). This focus can be explained, in part, by the rise of combative social media, which reverberated with contradictory and, more often than not, dubious content across the political landscape (Sanders, 2015). Perhaps just as important was the media's tendency to focus on, and as significantly perpetuate, the drama and scandal that it provoked (Markova and Gillespie, 2008, 5; Childress, 2016).

But also fueling this distrust was the context in which the election took place. Pointing to the rise in violent crimes, heightened intolerance, fear of the "other," an overall sense of system betrayal, as well as a decline in social engagement, Americans are experiencing a crisis of

DOI: 10.4324/9781032721125-13

trust—that is to say, a loss of trust in both each other and their institutions (Putnam, 2000; Hosking, 2014). Recent political data support this contention. Consider, for instance, the results of a September 2013 Gallup Poll, in which a mere 46 percent of respondents said they trusted their public officials, while only 19 percent trusted their Congressional representatives (Hosking, 2014, 9). As telling, since the Supreme Court's 2022 decision on Roe vs. Wade, as well as its failures to challenge Trump's immunity claims, trust in the high court has plummeted.

Far less commented upon, but certainly as perplexing, is that Trump—the candidate most associated with falsehoods—was, during the 2016 election campaign, considered more trustworthy than Clinton. This finding came as a great surprise, especially to the political pundits who had believed, along with me, that Trump's outlandish behavior would surely give Clinton the final victory, and by a large margin. Understandably so! As philosopher Sissela Bok pointed out in 1995, truthfulness was traditionally so prized in Western cultures that even white lies were considered unacceptable (Bok, 1995, 1999). Certainly, this is no longer the case in the United States.

9.2 The Crisis of Trust

How do we explain this paradox? Have we witnessed a phase transition such that the standards for trustworthiness have greatly changed (Bok, 1999)? If so, what accounts for this evolution? Addressing these questions is of utmost importance (Misztal, 1996, 17–18; Makela and Townley, 2013). Summarizing the societal significance of trust, Hosking (2014) attests:

> Configurations of trust are as important as those of power. Trust and distrust are part of the deep grammar of any society. The way in which we relate to each other, trust or distrust each other, determines much of our social behavior. In order to take decisions and act in real life, we need trust in other people, in institutions, or simply in the future. (33)

Yet, today, many scholars foresee an emerging crisis of trust, attributable to a decline of traditional values in the context of a highly differentiated, and risk-prone, complex, global society. Most likely, the consequences will be very severe. As these scholars warn, in today's increasingly complex environment, trust is more important than ever before (Luhmann, 1979; Giddens, 1990). Trust is the font of social capital, the general store of reciprocity that supports cooperation and collaboration (Luhmann, 1979, 88; Putnam, 1994, 2000; Fukuyama, 1995, 7, 33; Misztal, 1996; Baron et al., 2000; Ostrom and Walker, 2003; Rothstein, 2005, 4, 12; Buchanan, 2007; Sunderland, 2007; Hosking, 2014). As importantly, trust reduces transaction costs, increasing the effectiveness of political institutions and the efficiency of the economy (Williamson, 1985; North 1991; Fukuyama, 1995; Rothstein, 2005). At the same time, trust enhances sociability and personal well-being by integrating individuals within the social order (Simmel, 1964; Durkheim, 1984; Gambetta, 1993, 224; Putnam, 1994; Putnam, 2000).

A crisis of trust is especially concerning today, given our contentious political environment. Under the circumstances, we must wonder what will become of our democracy based as it is on truthfulness and transparency (Tilly, 2005, 5, 11). The threat to our democratic way of life looms ever larger as we enter a new global era where social structures are breaking down, diversity is proliferating, risks are high, and uncertainty is rife (Luhmann 1979, 105; Giddens, 1990; Luce, 2017). As history has shown, while trust is often taken for granted in stable times, in periods of major social upheaval, trust becomes problematic

(North, 1991; Misztal, 1996, 79–80; Frevert, 2009). In fact, as philosophers as far back as Hobbes have attested and as history has witnessed over the years, it is typically under such circumstances that autocratic, charismatic leaders take charge (North, 1991; Misztal, 1996, 36–37; Anderson, 2003; Frevert, 2009; Hosking, 2014, 21–22). Perhaps it is precisely such a situation that, today, explains Trump's, as well as that of other autocratic leaders', deepening appeal.

9.3   Defining Trust

To anticipate how the present-day "crisis of trust" might impinge on our political future—not to mention our everyday lives—we need a clearer idea of what we mean by trust and the role it plays in sustaining our social order. Unfortunately, even though trust is routinely referenced, it is only recently that the subject has been addressed in depth (Luhmann, 1979, 94; Misztal, 1996, 8–9, 22; Baron et al., 2000; Sunderland, 2007, 8–9).

Conceptualizing trust, it turns out, is no easy task. One problem is its multifaceted nature (Luhmann, 1979). Because trust is an aspect of all facets of life, it appears in multiple guises (Baron et al., 2000, 15; Markova and Gillespie, 2008; Makela and Townley, 2013). Trust is also situational: What accounts for trust in one set of circumstances, and for one type of person, may generate distrust in another (Ostrom and Walker, 2003, xx; Sunderland, 2007; Markova, 2008, xviii; Hosking, 2014). As problematic, because scholars study trust from diverse disciplines, they often perceive it through disparate lenses (Blomqvist, 1997). Hence, there are not only many different characterizations of trust; these are often at odds with one another (Misztal, 1996, 22; Makela and Townley, 2013; Hosking, 2014).

Consider, for example, the following perspectives. Starting from a position of methodological individualism, some scholars conceive of trust as *a rational calculation* about the likelihood that a goal will be met, based on the perceived incentives of those upon whom one is interdependent (Hardin, 2001). In contrast, others emphasize the *moral aspect of trust*, arguing that it is either a product of shared cultural experiences (Fukuyama, 1995), or some form of *emotional bonding* (Markova and Gillespie, 2008; Frevert, 2009; Grossen and Orvig, 2014). Still, others emphasize the *relational features of trust*, which are derived over time from repeated, reciprocal interactions (Putnam, 1994 56; Fukuyama, 1995, 7, 33; Baron, et. al., 2000; Ostrom and Walker, 2003; Rothstein, 2005; Tilly, 2005). *Sociocultural approaches* emphasize that trust is not only the product of negotiated, or dialogic, interpersonal communications but also of the specific social context in which such interaction takes place (Markova and Gillespie, 2008; Linell and Markova, 2013). Alternatively, some accounts minimize the need for interpersonal trust in today's world, emphasizing instead the role of formal institutions and governance rules in reducing risk (Fox, 1974, 67; Greif, 2006). Finally, technology advocates contend that trust can be automated by building it into software, as in the case of blockchain technology (Wright and Di Fillipi, 2015).

Notwithstanding these wide-ranging accounts, all of them are, in a critical sense, incomplete. Generalizing, we can say that trusting entails taking action to accomplish a goal in the face of some risk, in the belief that some "other"—be it a person, group, or institution—is competent and willing to help carry out that goal (Misztal, 1996, 34–35; Hardin, 2001; Ostrom and Walker, 2003). However, left unanswered and under-theorized is a generic explanation of trustworthiness, that is to say, the criteria that account for trust, and how these criteria evolve over time and in diverse settings (Hosking, 2014, 35–36).

In fact, in most accounts, trustworthiness cannot be generalized; instead, trust is ascribed to discrete variables that reflect a specific situation, which is typically related to a particular paradigm of trust. Hence, rational actor theorists explain trustworthiness in terms of cognitive criteria; moralists base trustworthiness on shared ethical values; relational theorists measure trustworthiness in terms of the quality of interactions, while neuroscientists suggest that it is a function, in part, of individuals' brain structures and personal histories (Luhmann, 1979; Rothstein, 2005).

Clearly, trust is a complex phenomenon. However, without a common reference point for what constitutes trustworthiness, we cannot compare and contrast the evolution of trust over time, nor garner insights about, and possible remedies for, the ostensible "crisis of trust" that we face as a nation today (Grossen and Orvig, 2014).

9.4 Can Standards Provide a Clue?

Perhaps our understanding of standards, and standard-setting processes, can help us establish a more abstract, and hence theoretical, account of trustworthiness and how trust emerges. As we shall see, in all instances of uncertainty and risk, it is standards, in whatever their guise, that engender confidence and an inclination to trust. Equally important, when standards are absent or disregarded, uncertainty mounts and distrust ensues.

9.4.1   Standards as the Focus of Analysis

What is it about standards that make them an appropriate unit for analyzing trust and trustworthiness? Most notably, standards, like trust, are essential for interaction. They are the interfaces that link people, people and things, and things and things. They connect diverse entities by providing the rules, or protocols, to be followed in order for objects to communicate and interact. More specifically, they govern the modes of interactions, define the conditions under which interactions can take place, and/or signify the appropriateness of interactions. To these ends, standards embody codes such as computer algorithms, signs, symbols, or social stereotypes that reduce the amount of information needed to engage under conditions of uncertainty (Misztal, 1996, 136). As Jaan Valsiner describes, these are the social representations of trust (Valsiner, 2008, xii).

9.4.2   Fostering Trust

When people interact, and adhere to the same standards, they are more inclined to both be trustworthy and to trust. By reducing uncertainty and providing assurances as to outcomes, standards bridge the gap between the known and the unknown, the present and the future (Valsiner, 2008, xiii). In this way, standards serve both as criteria of trustworthiness as well as a font of greater trust (Hosking, 2014).

One way to conceptualize the relationship between standards and trust is to think once again in terms of platforms (Johnson, 2005; Arthur, 2009; Gawer, 2009). Standards do not exist in isolation; rather, they are linked together in a network of complementary standards, which form platforms—one might say infrastructures—that link together and facilitate social, political, and economic activities (Johnson, 2005; Kauffman, 2008). This platform and the architecture that defines it determine the nature of relationships and hence the structure of society (Grewal, 2008).

9.4.3 Platforms of Trust

As noted above, by providing consistent and coherent rules, norms, and protocols, platforms that are built upon standards of interaction reduce uncertainty and engendering trust relations (Luhmann, 1979). As more people interact based on common standards, these trust relations become institutionalized in "nodal points of trust" (Markova and Gillespie, 2008, xix; Hosking, 2014, 56). When linked together atop the standards infrastructure, these nodal points constitute a higher-level configuration of "trust networks" (Tilly, 2005).

It is important to note, however, that not all societal networks are trust networks. Recall that common to the definitions of trust is the notion that trust entails some risk in achieving a high-stakes goal, as well as the confidence that some "other" will assist in accomplishing it. Similarly, we can differentiate trust networks from other interpersonal networks because their members "place their major collective enterprises at risk to malfeasance, mistakes, or failure by other members of the same networks" (Tilly, 2005, 5–6).

As telling, trust networks typically exhibit an architecture that is defined by strong ties, as opposed to the relatively weak ties associated with other types of social networks (Granovetter, 1983; Rogers, 2003; Burt, 2005; Borgatti et al., 2009; Easley and Kleinberg, 2010). Strong ties give rise to densely connected clusters, which comprise like-minded people. Their members tend to generate a common identity, which helps build solidarity and trust (Uzzi, 1996; Coleman, 1998; Burt, 2005). Moreover, because, in dense clusters, information and influence can travel rapidly along multiple paths, members of such groups can quickly identify cheaters and thereby limit opportunism and enhance trust (Krackhardt, 1992; Uzzi, 1996).

9.4.4 The Construction of Trust Networks

How do standardized trust networks come about, and how do they evolve over time? Because these networks are socially constructed, they are based upon, and adapted to, the circumstances at hand. Network standards emerge either organically from the bottom up or are prescribed from the top down by some established authority. Once in place, however, they create a positive feedback loop of cumulative causation. The benefits that accrue from following standards generate greater usage of the standards, which enhances their value and assures even greater adherence to them (Arthur, 1983; Rolfs, 2001; Grewal, 2008, 35–36).

Although standards platforms are relatively stable, they are subject to abrupt transformations. When social conditions undergo major structural changes, the value of standards can rapidly decline, in a phase transition (Pak, 1996; Holling, 2001; Beinhocker, 2008). New standard solutions are required if trust networks are to be maintained.

9.5 Trading Networks and Their Evolution

Trust was essential to the growth of long-distance trade. Avner Greif's account of the Maghribi traders, who operated in the Mediterranean during the early Middle Ages, provides a good example of a successful trade network based on trust. Organized exclusively in networks of shared norms, the Maghribis were highly successful so long as their interactions were limited to other traders with the same background. When, years later, trade expanded beyond their Mediterranean community, the Maghribs were overtaken by the Genoese, whose vertical trade networks mirrored a much more extensive trading area.

By incorporating foreigners, the Genovese were able to expand their trading networks beyond their own region, greatly increasing their economies of scale and scope. Significantly, instead of cultural mores, the Genovese employed formal institutions to maintain trust (Greif, 2006).

The *guanxi* networks in China provide a similar example of how societal changes might generate new types of networks at the expense of previous ones. The coming of the Communist regime in China made it impossible to maintain traditional family networks, which previously had provided a source of trust. *Guanxi* networks, which are based on influence and interpersonal relations emerged to assure the availability of resources in a planned economy (Lui, 2008, 62). Today, these networks serve a similar purpose in the global capitalist economy, as global companies have learned that to develop trading relationships with China, they must associate themselves with *guanxi* networks (Lui, 2008, 62–64).

Notwithstanding the benefits of standards, and the trust networks they engender, standards, as we have seen, are far from neutral. Determining behavior across a wide range of activities, they constitute major sources of power (Grewal, 2008, 13–15; Hakli and Minca, 2016). Opting out of such networks is costly, as *guanxi* networks illustrate. By doing so, one foregoes access to the benefits of trust and trusting relationships. Moreover, while standards and trust may foster collaboration within a network, they may also instill fear by discriminating against, and excluding those on the outside (Buchanan, 2007; Tilly and Tarrow, 2017).

9.6 Standards of Trust in the Victorian Era

Can standards generate trust while simultaneously reinforcing exclusion and hierarchical power relations? The class system in Victorian England provides a clue. When novel social conditions emerged in industrial England, new standards were required to maintain effective trust relations (Markova and Gillespie, 2008). These standards not only reflected but also reinforced the existing social structure.

Prior to the industrial era in England, trust was mostly taken for granted. People typically adhered to a common set of moral standards, which were thought to be innate in peoples' minds. Believing people were inherently trustworthy, they generally behaved accordingly. Hence, trust was not viewed as a major concern (Misztal, 1996, 37–40; Himmelfarb, 2005, 39; Evensky, 2011). However, with the rise of industrialization, accompanied as it was by an increase in specialization and the division of labor, universal norms declined. As people were increasingly divided based on their diverse experiences and interests, a more well-defined class structure emerged, one that was shaped and reinforced by a new regimen of standards. In contrast to the universal norms of the Enlightenment era, these standards were segmented and particularized according to class distinctions. Designed to control relationships both between and among classes, these standards focused on the attribution of personal characteristics such as codes of etiquette, mores of respectability, sociability, and reputation (Simmel, 1964; Sunderland, 2007; Markova and Gillespie, 2008, 14, 16).

9.6.1 Standard for the Working Class

Consider how standards supported and regulated the lives of the working classes. While early industrial standards enhanced trust and social capital within the sphere of workers, they also regularized and bounded relationships beyond them. Establishing trusting relations was especially important for working people because, having limited resources, they depended on their neighbors, friends, and relatives to help in times of trouble (Sunderland, 2007, 87).

By adhering to norms established from the bottom up within their own neighborhoods, workers demonstrated and communicated their trustworthiness. Most important was their frequent participation in community associations and events that supported reciprocity and provided mutual support. Key venues were neighborhood associations, friendly societies, sporting contests, band concerts, and parades. Equally important were regular visits to local pubs, where workers gathered to gossip, share their stories, and solidify their common identities. Drinking signaled trustworthiness, whereas sobriety was a cause for suspicion (Sunderland, 2007, 60–62). Because one's reputation was linked to that of the community, everyone upheld these standards. Neighborhood norms were maintained, and reputations were made or destroyed, through active gossip channels (Sunderland, 2007, 64).

9.6.2 Standards for the Middle Class

Aiming to move up the social hierarchy, middle-class members differentiated themselves by their "virtuous" living. In contrast to the spirited style of the working class, the middle class signaled their trustworthiness with norms of rectitude. To demonstrate their uprightness and develop a reputation for honesty and straightforwardness, they adhered to a strict code of etiquette (Sunderland, 2007, 32–39). Thus, they measured their speech to conceal their emotions. Moreover, they projected an air of respectability by working hard, remaining sober, attending church, and joining reputable associations. The middle classes also dressed appropriately for each occasion based on a common set of standards. Likewise, household objects, such as heavy, polished furniture, and pictures of ancestors, were displayed conspicuously to indicate their status. In contrast to the working class, where standards were maintained and reinforced in pubs, the middle classes promoted social coherence through gift exchanges, domestic socials, and interactions at concert halls, art galleries, and museums (Sunderland, 2007, 51). When rules were breached, offenders were disciplined through gossip, insults, and exclusion (Sunderland, 2007, 37–38).

9.6.3 Standards for the Aristocracy

In a rapidly changing industrial society, upper-class status became more precarious, as the opportunities for the upper middle classes to rise in status were expanding. To shore up their positions and to solidify trust in each other and their way of life, the aristocracy differentiated itself by setting a standards barrier around its space.

The *calling card* was at the centerpiece of aristocratic life. These cards not only allowed people to enter the elite social circle but also kept the unwanted out (Hoppe, nd). The designs and formats of these cards were also standardized, signally the status of the user. As well, how and by whom calling cards were to be distributed and the circumstances under which they were to be delivered were also standardized. Likewise, whether and how the cards were received regulated future interactions and determined the status relationships among the various parties (Hoppe, nd). In effect, calling cards standardized aristocratic life.

Equally important in reinforcing the upper class's high status was its common lifestyle, which was expressed and telegraphed via their sumptuous possessions and personal effects as well as their everyday performance. Conformity was the rule, as deviant behavior was severely punished. To foster compliance, publishers disseminated popular guidebooks describing standards of courtesy and etiquette, including descriptions of the appropriate poses to take when receiving a letter, greeting an acquaintance, performing the minuet, etc. (Morgan, 1994; Nivelon, 2003).

9.6.4 Standards and the Social System

Industrial England illustrates not only the symbiotic relationship between standards and trust but also the link between standards, trust, and the social environment. With the onset of industrialization, new norms emerged that corresponded to, and reinforced, a highly differentiated three-tiered class structure. At the micro level, these standards divided people according to their social status. At the same time, however, these same standards supported cooperative relations at the macro level among all interdependent actors (Durkheim, 1984). By providing a modicum of confidence and assurance about future relationships and outcomes, Victorian Age standards fostered productive interactions that generated externalities in the form of trust networks, which both reinforced and cut across class divisions (Misztal, 1996, 29).

9.7 Today's Standards Gap

Much as in the industrial age, societies today are witnessing structural changes that threaten the normative foundations and institutional frameworks that fostered trust throughout the post-war era (Biersteker, 1995; Arrighi and Silver, 1999; Freeman and Louca, 2001; Perez, 2010; Luce, 2017). One might ask, then, does trust have a future? To address this question, we must look at interactions—their structure and quality—as well as at the standards that define and support them. Trust formation is a product of communication and interaction via an infrastructure of standards and standardization.

In contrast to the Victorian era when standards supported interactions and association both within and across group boundaries, in today's post-modern era, traditional standards are being challenged, while barriers to their replacement are being set in place. As Misztal (1996) opines:

> The collapse of traditional standards around such issues as family, work, discipline, the decline of industrial and class identities, the increase of culturally specific identities (ethnic, rational, territorial), the weakening of the welfare state and the decline of national boundaries, all raise a vital question for modern society; where are we to look for social solidarity, cooperation, and consensus? (11)

9.8 Distantiation and Communication Technologies

One major force driving these developments is the unprecedented advance in communication and information technologies. As sociologist Anthony Giddens has pointed out, by extending communication far and wide, these technologies have uncoupled the relationships between time and space, and space and place, so that uncertainty looms ever larger (Luhmann, 1979; Giddens, 1990). As problematic, given the stretching of time and space, opportunities to establish standards of trust via platforms of interpersonal interactions and associational life are greatly diminished (Luhmann, 1979; Giddens, 1990).

9.8.1 Workplace Distantiation

One need only consider today's workplace where face-to-face interactions are increasingly held to a minimum in the name of efficiency. Instead of sharing ideas and reinforcing common standards in hallways, or at the watercooler, workers are isolated from one another, communicating online with diverse locals (Brown and Duguid, 2017, 108–109). Working independently on short-term gigs, or miles apart from their traditional workspaces, workers have little opportunity to bond. At best, their networks consist of weak ties,[1] which are not conducive to building trust.

9.8.2 Marketplace Distantiation

Similarly, disintermediation is occurring in the market. With the advent of electronic markets, buyers and sellers are increasingly dis-embedded from their social contexts. In many supermarkets, for example, self-help checkout stations have deprived individuals of even the most minimal social contact in their shopping experiences. Moreover, as computer algorithms replace trusting relationships, online buyers increasingly resemble the autonomous, economic man posited by neoclassical economists, whose behavior is governed solely by achieving the highest value at the lowest possible price (Ariely, 2008; Beinhocker, 2008).

9.8.3 Rural–Urban Distantiation

Advanced communication and information technologies have also contributed to the separation of rural and urban regions, helping to trigger many of the ideological cleavages we witness today. With the expansion of the global economy, massive industrial urban regions have emerged as the center of today's economic activities. Thus, today, 40 percent of the US labor force works in counties that together make up only 1.5 percent of the US land mass (Scott and Storper, 2010). The result has been the emergence of a hierarchical structure of global cities and their city regions (Storper and Scott, 1997; Sassen, 2001). Trending as well is the decreased proximity of former city dwellers, as those of low-income people have been forced to move to the suburbs and beyond to find affordable living (Luce, 2017; Florida, 2017).

9.8.4 Distantiation in the Media

Distantiation in the media has also led to declining faith in the free press, a cornerstone of our democracy (Vincent, 2018). Recall the cartoon, "On the Internet nobody knows you're a dog." In the early days of the Internet, this quip reflected a presumed victory for privacy and anonymity, as several academics championed the idea that the Internet would liberate people from excessive conventions and constraints (Rheingold, 1993, 21). What they failed to recognize, however, was the trade-off between the benefit of anonymity and privacy and the loss of accountability and trust. Absent interactions associated with identifiable behaviors, standards of trustworthiness, and the trust associated with them will fail to emerge. Not surprisingly, during the 2016 election, the voting public was unable to ascertain who was, and who was not, a Russian.

9.9   Technological Solutions

Given distantiation, many look to technology solutions, such as blockchain technology, to enhance trust (Antonopoulos, 2014; Atzori, 2015; Wright and De Filippi, 2015). A blockchain is a distributed database comprising records of all transactions, much like a collective bill of lading. What allows the blockchain to reduce uncertainty and enhance trusting relationships is that it is populated by encrypted data, which not only can be automatically recorded and verified in all network nodes but is also irreversible and thus resistant to human error and tampering (Antonopoulos, 2014; Atzori, 2015, 2; Wright and DeFelippi, 2015). In this fashion, the blockchain provides trust via calculation, shifting the focus of trustworthiness from people to mathematic algorithms (Antonopoulos, 2014).

While blockchain technology can provide greater security and accuracy for transactional data, it is extremely limited in supporting the myriad aspects of trust and trustworthiness. Because blockchain technologies, such as cryptocurrencies, are used to support illegal activities, they may very well reinforce apprehension at the societal level. Also, because trust building requires a modicum of interpersonal interactions, blockchain technologies cannot generate the social capital required to bind individuals together and develop long-lasting trust among them.

9.10   The Future of Trust

This state of affairs does not bode well for the future of trust (Edelman Trust Survey 2017). As Luhmann and others have emphasized, the shift from trust to fear can occur precipitously, as in an epidemic, or phase transition (Luhmann, 1979). Moreover, once fear has taken hold, trust rarely stages a comeback (Rothstein, 2005, 18–21; Hosking, 2014, 29). Making matter worse, much as we are witnessing today, autocratic political leaders reinforce the lack of trust by exploiting fears and inducing uncertainty as a means of gaining inordinate power (Markova and Gillespie, 2008, 3; Rothstein, 2005, 13, 18). As one might suspect, one of their primary strategies is to undermine societal standards and restrict the interactions and communications that engender them.

The Bolshevik strategy in post-revolutionary Russia provides a case in point. As Hosking (2014) points out, the new Soviet state was deliberate in its efforts to undermine everyday norms and standards (Hosking, 2014, 21–22). Moreover, the regime sabotaged the legitimacy of traditional authorities and institutions including the czar, the church, the family, and property rights. Uncertainty abounded as a result so that fear rather than trust governed interactions, making meaningful communication nigh impossible. Instead of relying on one another, Soviet citizens had only one fallback—to rely on the State (Hosking, 2014).

Today, it is most alarming that when contemplating a second Trump presidency, shades of Bolshevism come to mind. As described by Livitsky and Ziblatt (2019), Trump is a "serial norm breaker." From the unbridled insults he hurls at his competitors and adversaries to his attacks on democratic institutions—the rule of law and even the Constitution—Trump seeks to generate chaos, instill fear, and undermine the bonds of trust (Tobin, 2018). To this end, he has doctored the truth, replacing facts with his own distorted reality. His antics, moreover, have not been limited to the domestic realm. Reversing decades of foreign policy, he has renounced existing agreements and lavished praise on loathsome dictators while heaping scorn on the country's most loyal democratic allies (Calamur, 2016; Rose, 2017). Like a chameleon, lying and shifting his positions, Trump cannot be pinned down, nor can he be held accountable. Deeply offended in the face of criticism, Trump lashes out in erratic, and at times nonsensical, "tweet storms," raging against the limits of his power.

One cannot but wonder what accounts for Trump's enduring appeal and his 2024 presidential victory. Notwithstanding his 91 indictments, a majority of Americans have continued to support him in the hope that by doing so, they might better promote their own power, agendas, and personal needs. One need only consider, for example, Jeff Bezos' decision not to publish *the Washington Post*'s Editorial Board's endorsement of Kamala Harris. Bezos' decision brings to mind the same misguided strategy that many right-wing Germans and Italians pursued in the Thirties, decisions they came to regret during the rise of Hitler and Mussolini.

In explaining Trump's recent electoral success, it is important to keep in mind Max Weber's differentiation between leadership types as they relate to the societal contexts in which leaders rule. In his seminal work, *The Theory of Social and Economic Organization* (1978), Weber identifies three "ideal types" of leadership: traditional leaders, whose authority is based on customs; legal/rational leaders, whose authority is centered on bureaucratic government and the rule of law; and charismatic leaders, whose authority is derived from the strength, and appeal, of the leader's individual personality. Weber adds that in each case, leadership will only prevail, if there is a high level of congruence, and hence trust, between the "standardized" role leaders play and the expectations of their followers. Moreover, as he notes, the substance of these interrelated roles and the subsequent relationships they engender are contingent, to a large extent, on the prevailing situations and circumstances in which interactions take place.

Donald Trump may not have read Max Weber, but he certainly understands the dynamic that has been playing out between his personality and the expectations of his followers.[2] Viewing himself as a charismatic leader, he has—much to the chagrin of his advisors—avoided concrete policy discussions in favor of campaigns based on his "strongman" personality, a strategy that aligned with the electorate's expectations and proved key to his success. To gain the presidency, he has not only structured the situation in his favor, by stoking the flames of hatred and fear but also created a persona and corresponding role that, under the circumstances, matches the needs of his constituents. Much like Mussolini, Trump has molded people's desires to conform to his own needs and then offered himself up as the only person capable of meeting their expectations.

Equally important in explaining Trump's hold on the electorate is the "Trumpian" movement and networked platforms, such as Fox News and Twitter (X), where his supporters coalesce to promote conspiracy theories—such as people eating cats and dogs—and cultural memes that reinforce Trump's charismatic image. As Randall Collins might say, these platforms provide a common space to carry out "interaction rituals," where shared symbols generate intense mutual focus, and compelling emotional experience among the participants (Collins, 2004, 12–14). These symbols play a major role not only in unifying and reinforcing strong tie connections but also in enhancing the status of Trump's followers. Thus, we see the flourishing of MAGA hats, Trump trading cards, Trump Bibles, and Trump sneakers. Given the intensity of these network interactions, both with people and with things, it is costly for those who have been entrained in Trump's networks—to let go (Grewal, 2008). What is at stake for them is not only the connections to like-minded people but also, as importantly, the emotional rewards to which these connections give rise.

What is required to stem the momentum fomented by Trump's election to the presidency? According to post-election pundits, opponents of Trump, and the movement he stands for, must engage more fully with the wide array of disillusioned voters, refocusing their messages to better address voters' concerns. Such may very well be the case when viewing the situation from a rational, individualist perspective. But this is neither the whole story nor the long-term solution. One need only consider the fact that Kalama Harris, unlike Trump, did propose specific policies designed to address many of the economic needs of the lower and middle classes. Nonetheless, and notwithstanding the unprecedented voter enthusiasm she inspired, Harris was perceived by a majority of the voters as being—in Max Weber's terminology—a bureaucratic/legalistic leader. In other words, Harris failed because, in a time of extreme crisis, brought about to a considerable degree, by Trump's machinations—her role performance did not match a majority of the electorate's role expectations.

Keep in mind, however, this conclusion is not the sole story, nor—hopefully—the end of it. When considering the real threat of power-wilding autocrats from a long-term perspective, and in the context of complexity theory, we see that what is needed to alter the present scenario is a multidimensional strategy, one that circumvents the national political arena to a considerable degree, so that we might, working from the bottom up, regenerate trust and the collaborative standards that foster it. In this way, we may be able to fill the gaps left by the stretching of time and space, which unethical politicians, and other power brokers, have filled in, much to their own advantage.

Notes

1 See Chapter 4 for a discussion of weak ties and their impacts.
2 See, for example, the discussion in Chapter 10.

References

Anderson, Christopher (2003) "Hobbes, Locke, and Hume on Trust and the Education of the Passions,"*The New England Journal of Political Science* 1 (1), 52–80.

Antonopoulos, A. (2014) "Bitcoin Security Model: Trust by Computation," O'Reily-Radar. http://radar.oreilly.com/2014/02/bitcoin-security-medel-trust-bycomputation.html

Ariely, Dan (2008) *Predictably Irrational*, New York, NY: Harper Collins.

Arthur, Brian (1983) *On Competing Technologies and Historical Small Events: The Dynamics of Choice under Increasing Returns*, Mimeo, Stanford University.

Arthur, Brian (2009) *The Nature of Technology: What It Is and How It Works*, New York, NY: Free Press.

Arrighi, Giovanni, and Beverly Silver (1999) *Chaos and Governance in the Modern World System*, Minneapolis, MN: University of Minnesota Press.

Atzori, Marcella (2015) Blockchain Technology and Decentralized Governance; Is the State Still Necessary. http://ssrn.com/abstract=2731132

Baron, Stephen, John Field, and Tom Schuller (2000) *Social Capital, Critical Perspectives*, New York, NY: Oxford University Press.

Beinhocker, Eric (2008) *The Origin of Wealth: Evolution, Complexity, and the Radical Remaking of Economics*, Cambridge, MA: Harvard Business School Press.

Biersteker, Thomas J. (1995) "The Triumph of Liberal Economic Ideas in the Developing World," in Stallings, Barbara, ed. *Global Change, Regional Response in the Developing World, The New International Context of Development*, New York, NY: Cambridge University Press.

Blomqvist, Kirismarja (1997) "The Many Faces of Trust," *Scandinavian Journal of Management* 13 (3), 271–286.

Bok, Sissela (1995) "Donald Trump and the Culture of Lying," *New Yorker political scene podcast. http://www.newyorker.com/podcast/political-scene/donald-trump-and-the-culture-of-lying*

Bok, Sissela (1999) *Lying: Moral Choice in Public and Private Life*, New York, NY: Vintage Books.

Borgatti, et al. (2009) "Network Analysis in the Social Sciences," *Science*, 323 (5916), 892–895.

Brown, John Seely, and Paul Duguid (2017) *The Social Life of Information*, Boston, MA: Harvard Business School Press.

Buchanan, Mark (2007) *The Social Atom: Why the Rich Get Richer, Cheaters Get Caught, and Your Neighbor Usually Looks Like You*, New York, NY: Bloomsbury.

Burt, Ron (2005) *Brokerage and Closure: An Introduction to Social Capital*, New York, NY: Oxford University Press.

Calamur, Krishnadev (2016), "Nato Schmato," *The Atlantic*.

Chan, Melissa (2016) "Donald Trump More Trustworthy Than Hillary Clinton, Poll Finds." https://www.theatlantic.com/politics/archive/2016/09/clinton-trust-sexism/500489/

Childress, Sarah (2016), "Study: Election Coverage Skewed by Journalistic Bias." http://www.pbs.org/wgbh/frontline/article/study-election-coverage-skewed-by-journalistic-bias/

Coleman, James (1998) *Foundations of Social Theory*, Cambridge, MA: Belnap Press.

Collins, Randall (2004) *Interaction Ritual Chains*, Princeton, NJ: Princeton University Press.

Durhally, Lena Aberdene (2016) "Why Trust Matters in the Election," *Huffington Post*. http://www.huffingtonpost.com/lena-aburdene-derhally/why-trust-matters-in-the-_b_9366228.html

Durkheim, Emile (1984) *The Division of Labor in Society*, New York, NY: The Free Press.

Easley, David, and Jon Kleinberg (2010) *Networks, Crowds and Markets: Reasoning about a Highly Connected World*, New York, NY: Cambridge University Press.

Edelman Trust Survey (2017) http://www.edelman.com/trust2017/

Evensky, Jerry M. (2011)"Adam Smith's Essentials: On Trust, Faith, and Free Markets, *Economics Faculty Scholarship Paperm Paper 6*. http://surface.syr.edu/ecn/6

Florida, Richard (2017) *The New Urban Crisis*, New York, NY: Basic Books.

Fox, Alan (1974) *Beyond Contract: Work, Power and Trust Relations*, London: UK; Faber and Faber.

Freeman, C., and F. Louca (2001) *As Time Goes By: From the Industrial Revolution to the Information Revolution*, New York, NY: Oxford University Press.

Frevert, Ute (2009) "Does Trust Have a History? Max Weber Lecture No. 2009/01, San Domenico di Fiesole, Italy.

Fukuyama, Francis (1995) *Trust: The Social Virtues and the Creation Of*, New York, NY: The Free Press.

Gawer, Annabelle (2009) *Platforms, Markets and Innovations*, Northampton, MA: Edward Elgar.

Gearan, Anne (2016) "Can Hillary Clinton overcome her trust problem?" *The Washington Post*. https://www.washingtonpost.com/politics/can-hillary-clinton-overcome-her-trust-problem/2016/07/03/b12eeb52-3fd8-11e6-84e8-1580c7db5275_story.html?utm_term=.afe3c0dbb61d

Giddens, Anthony (1990) *The Consequences of Modernity*, Stanford, CA: Stanford University Press.

Granovetter, Mark (1983) "The Strength of Weak Ties: A Network Theory Revisited," *Sociological Theory* 1, 201–233.

Greif, Avner (2006) *Institutions and the Path to the Modern Economy: Lessons from Medieval Trade*, New York, NY: Cambridge University Press.

Grewal, David Singh (2008) *Network Power: The Social Dynamics of Globalization*, New Haven, CT: Yale University Press.

Grossen, Michele, and Anne Salazar Orvig (2014) "Dialogical Approaches to Trust in Communication," in Linell, Per, and Ivana Markova, eds. *Dialogical Approaches to Trust in Communication*, Charlotte, NC: Information Age Publishing, Inc.

Hakli, Jouni, and Claudio Minca (2016) *Social Capital and Urban Networks of Trust*, New York, NY: Routledge.

Hardin, Russell (2001) "Conceptions and Explanations of Trust," in Cook, Karen S., ed. *Trust in Society*, New York, NY: Russell Sage Foundation.

Himmelfarb, Gertrude (2005) *The Roads to Modernity, The British, French, and American Enlightenments*, New York, NY: Vintage Books.

Holling, C.S. (2001) "Understanding the Complexity of Economic, Ecological, and Social Systems," *Ecosystems* 4, 390–405.

Hoppe, Michelle (nd) "Calling Cards and the Etiquette of Paying Calls." http://www.literary-liaisons.com/article026.html

Hosking, Geoffrey (2014) *Trust: A History*, Oxford: UK: Oxford University Press.

Johnson, Steven (2005) *Where Good Ideas Come From: The Natural History of Innovation*, New York, NY: Riverhead Books.

Kauffman, Stuart (2008) *Reinventing the Sacred*, New York, NY: Basic Books. checker/wp/2017/02/12/stephen-millers-bushels-of-pinocchios-for-false-voter-fraud-claims/?utm_term=.9017ff0dd6b1

Krackhardt, D. (1992) "The Strength of Strong Ties: The Importance of Philos in Organizations," in Nohria, N., and R. Eccles, eds. *Networks and Organizations: Structure, Form and Actions*, Boston, MA: Harvard Business School Press.

Lui, Li (2008) "Trust in China," in Markova, Ivana, and Gillespie Alex, eds. *Trust and Distrust: Sociocultural Perspectives*, Charlotte, NC: Information Age Publishing, Inc.

Linell, Per, and Ivana Markova, eds. (2013), *Dialogical Approaches to Trust in Communications*, Charlotte, NC: Information Age Publishing, Inc.

Livitsky, Steven and Danial Ziblatt (2019) *How Democracies Die*, New York, NY: Crown.

Luce, Edward (2017) *The Retreat of Western Liberalism*, New York, NY: Little Brown.

Luhmann, Niklas (1979) *Trust and Power*, Chichester: John Wiley & Sons.

Makela, Pekka, and Cynthia Townley, eds. (2013) "Introduction," *Trust: Analytic and Applied Perspectives*, New York, NY: Vintage Books.

Markova, Ivana, and Alex Gillespie (2008) *Trust and Distrust: Sociocultural Perspectives*, Charlotte, NC: Information Age Publishing, INC.

Misztal, Barbara (1996) *Trust in Modern Societies, The Search for Social Order*, New York, NY: Blackwell Publishers Inc.

Morgan, Marjorie (1994) *Manners, Morals and Class in England, 1774–1858*, London, UK: Macmillan Press LTD.

Nivelon, Francois (2003) *The Rudiments of Genteel Behavior: Facsimile Reprint of the First Edition of 1737*, London, UK: Paul Holberton Publishing.

North, Douglass (1991) *Institutions, Institutional Change, and Economic Performance*, New York, NY: Cambridge University Press.

Ostrom, Elinor, and James Walker, eds. (2003) *Trust and Reciprocity: Interdisciplinary Lessons from Experimental Research*, New York, NY: Russell Sage Foundation.

Pak, Per (1996) *How Science Works, The Science of Self-Organized Criticality*, New York, NY: Copernicus/Springer.

Perez, Carlotta (2010) "Technological Revolutions and Techno-Economic Paradigms," *Cambridge Journal of Economics* 34 (1), 185–202.

Putnam, Robert D. (1994) *Making Democracy Work: Civic Traditions in Modern Italy*, Princeton, NJ: Princeton University Press.

Putnam, Robert D. (2000) *Bowling Alone: The Collapse and Revival of American Community*, New York, NY: Simon and Schuster.

Rheingold, Howard (1993) *The Virtual Community: Homesteading on the Electronic Frontier*, Reading, MA: Addison Wesley.

Rolfs, J. (2001) *Bandwagon Effects in High Technology Industries*, Cambridge, MA: MIT Press.

Roberts, Dan (2016), "Why Hillary Clinton lost the election: the economy, trust and weak message," *The Guardian*.

Rogers, Everett (2003) *Diffusion of Innovations*, 5th edition, New York, NY: Free Press.

Rose, Gideon (2017) *Trump and the Allies, Foreign Affairs*.

Rothstein, Bo (2005) *Social Traps and the Problem of Trust*, Cambridge, UK: Cambridge University Press.

Sanders, Sam (2015), "Did Social Media Ruin Election 2016." http://www.npr.org/2016/11/08/500686320/did-social-media-ruin-election-2016.

Sassen, Saskia (2001) *The Global City: New York, London, Tokyo*, Princeton, NJ: Princeton University Press.

Scott, Allen, and Michael Storper (2010) "Regions, Globalization, Development," *Regional Studies*, 37, 579–593.

Silver, Nate (2018) "How Much Did Russian Interference Affect the 2016 Election?" *FiveThirtyEight*. https://fivethirtyeight.com/features/how-much-did-russian-interference-affect-the-2016-election/

Simmel, Georg (1964) *The Sociology of Georg Simmel*, New York, NY: The Free Press.

Storper, Michael, and Allan Scott (1997) *The Regional World: Territorial Development in a Global Economy*, New York, NY: Guilford Press.

Sunderland, David (2007) *Social Capital, Trust, and the Industrial Revolution*, New York, NY: Routledge.

Tilly, Charles (2005) *Trust and Rule*, New York, NY: Cambridge University Press.

Tilly, Charles, and Sidney Tarrow (2017) *Contentious Politics*, 2nd edition, New York, NY: Oxford University Press.

Tobin, Jeffrey (2018) "Donald Trump and the Rule of Law," *The New Yorker*.

Uzzi, B. (1996) "The Sources and Consequences of Embeddedness for Economic Performance in Organizations: The Network Effect," *American Sociological Review* 61, 674–698.

Valsiner, Jaan (2008) "Introduction," in Markova, Ivana, and Alex Gillespie, eds. *Trust and Distrust: Sociocultural Perspectives*, Charlotte, NC: Information Age Publishing, Inc.

Vincent, James (2018) "Former Facebook exec says social media is ripping society apart." https://www.theverge.com/2017/12/11/16761016/former-facebook-exec-ripping-apart-society

Williamson, Oliver (1985) *The Economic Institutions of Capitalism*, New York, NY: Vintage Books.

Wright, Aaron, and Primavera De Filippi (2015) "Decentralized Block Chain Technology." http://ssrn.com/abstract=2580664

10

CRAFTING IDENTITY WITH STANDARD MEMES AND SYMBOLS

10.1 Introduction

I recall the time several years ago, when I was making a presentation on standard setting at a conference in Japan. It was my first trip to Japan, and I didn't quite know what to expect. To screw up my courage, and buffer my morale, I donned a brand-new outfit, decorated with feminine frills, and accented with a stylish pair of pumps. The speech went exceptionally well, and I returned home brimming with pride. Not long thereafter, I received some nice photographs of the event. What a surprise! Despite my efforts to portray a certain image of myself, the photos of me were labeled Mr. Garcia. Apparently, I had little control over my identity. How I was defined, it turned out, had nothing to do with my appearance and everything to do with the role I played as a "standards expert"—a role that, in Japan, is clearly associated with the masculine sex. The experience reminded me of an article I had read in *The New York Times*, written by Didi Kirsten Tatlow, and entitled, "In China, a Respected Ms. May Be Labeled Mr." (Tatlow, 2013, A1). It seems that, in today's Japan, female opinion leaders and scholars are referred to in masculine terms. In fact, such labels are considered a sign of great respect.

Thinking along these lines, I am reminded of Alice's encounter with the Caterpillar during her adventures in Wonderland. You might recall the instance as well.

The Caterpillar and Alice looked at each other for some time in silence: at last, the Caterpillar took the hookah out of its mouth, and addressed her in a languid, sleepy voice.

'Who are YOU?' said the Caterpillar. This was not an encouraging opening for a conversation.

Alice replied, rather shyly, "I hardly know, sir, just at present—at least I know who I was when I got up this morning, but I think I must have been changed several times since then."

'What do you mean by that?' said the Caterpillar sternly. 'Explain yourself!'

'I can't explain **MYSELF**, I'm afraid, sir' said Alice, 'because I am not myself, you see.'

DOI: 10.4324/9781032721125-14

'I don't see', said the Caterpillar.

'I am afraid I can't put it more clearly', Alice replied very politely, 'for I can't understand it myself to begin with; and being so many different sizes in a day is very confusing.'

(Carroll, 2009)

Believe it or not, Alice's identity problem was one of the standards. In that strange wacky Wonderland, Alice lacked stable and coherent standards against which to define herself. In today's terminology, we might say that Alice was experiencing an "identity crisis" (Erikson, 1968; Cote and Levine, 2002). Much as I had encountered during my trip to Japan, Alice became unsure about her own identity when she found herself uprooted and in a strange environment where there were no standards, and her physical persona was in a constant state of flux.

While we can easily understand the function that standards play in determining the identity of a thing, as for example in the case of a product description, it is somewhat more difficult to think about the role standards play in identifying people in standardized terms (Bowker et al., 2000). But, just as the components of an automobile need to be standardized so they can fit together and operate as a working vehicle, so individuals require standard identities if they are to be integrated into, and fully function within, a social order.

10.2   Standardized Roles

In fact, even in today's individualist culture, it is standard identities (conceived as roles) that provide the interfaces that allow us to interact and cooperate with one another in our complex environment. For, in any given context, individuals occupy positions that carry with them a complex set of role expectations defining obligations and values (Merton, 1949; Katz and Kahn, 1966; Biddle, 1986; Burke and Stets, 2009). As described by Burke and Stets (2009):

... the core of an identity is the categorization of the self as an occupant of a role. Once the incorporation into the self of meanings and expectation associated with that role and its performance, these expectations and meanings form a set of standards that guide behavior. (225)

According to psychological identity theories grounded in the works of Erik Erikson (1968), and sociological theories based on the symbolic interactionism of George Herbert Mead (1967) and Herbert Blumer (1986), individuals form their identities by taking on, and performing, roles from the repertoire that, in the context in which they are situated, are available to them. Viewed through the lens of complexity theory, we might call the sum of these roles emergent platforms and characterize them as small worlds.

Individuals recognize each other and develop expectations about each other, based on the identity roles that each plays (Cote and Livine, 2002). It is in this way that social structure is created, and trust is established, allowing coordination and collaboration to take place (Katz and Kahn, 1966). However, as described below, we must keep in mind that identity standards, as well as the lack thereof, are also the basis upon which discrimination and cultural wars take place (Buchanan, 2007).

10.3   Self-Identification and the Leeway for Self-Actualization

How much leeway do we have to define ourselves? The answer is complex, as I found out during my trip to Japan. Our freedom of choice depends not only on the specific context and the standardized roles that are available to us but also on the resources that an individual has—a point that Stryker and Serpe emphasize in their exposition of "structural symbolic interaction". This kind of capital not only consists of material and sociocultural resources; it also includes a resilient individual core, or psyche, that can adapt and take advantage of changing contexts and situations (Goffman, 1967; Cote and Levine, 2002; Burke and Stets, 2009, 4). But, as Thoit reminds us, these resources are distributed unevenly. As she notes:

> People's social status—gender, race, ethnicity, education, social class, age, religion and nationality—influences the opportunity they have for role-identity acquisition and accumulation. Regardless of motivation, skill, or preparation, due to their locations in the social structure, some people may be actively barred from entry into role domains, be kept unaware of the possibilities for entry, or be discouraged by others from trying.
>
> *(Burke et al., 2003, 199)*

10.3.1   Gaining Individual Agency through Role Performance

Erving Goffman has argued that individuals can gain the agency needed to define themselves and develop their individual platforms through their role performance. Although not formally associated with symbolic interactionism, Goffman's view of identity formation is in many ways consistent with it. In fact, Goffman used a theatrical metaphor to describe the process of symbolic interaction (Goffman, 1967). In performing certain roles, individuals can frame their image, as well as the situation, so as to gain acceptance from their audiences. Thus, in contrast to Mead and Blumer's emphasis on individuals *"taking on"* societal roles, Goffman claims that actors gain agency by consciously manipulating their performances through their appearances, demeanors, and behaviors to establish their identities and the roles they play (Goffman, 1967).

10.3.2   Trump—The Ultimate Role Performer

A prime illustration of Goffman's perspective is the performance of the President-elect, Donald Trump. Known to be a consummate grifter, Trump—even before his election—cast himself as a successful billionaire who didn't need to profit from being president. As he claimed, his only goal was to "Make America Great" again. Creating a heroic image of himself to fit this claim, Trump donned a red baseball hat, a yellow hair piece, and orange makeup. To project a picture of health and virility, he claimed to weigh only 2015 pounds. As well, to appear *macho,* Trump demeaned women and fueled rumors of his sexual exploits. And, all the while, he bolstered his appeal with white supremacist and evangelicals by behaving rudely towards foreigners and employing racial dog whistles.

How did Trump succeed in turning his image to such an advantage? In short, he created an individual platform drawing on tropes that mirrored the identities of his potential followers, linking them to him through role convergence and the force of charisma (see Weber, 1947). By displaying straight forwardness, strength, and purpose, he resonated with the growing number of Americans who, experiencing anomie, were in a state of despair (Durkheim, 1984). At his rallies, Trump spoke directly to their grievances, even generating some where there had been none before. It was through this process that Trump established an elaborate personal platform, a cult that centered on him. So total was the shared identity between Trump and his followers that, when calling on the mob to protest Biden's presidency by attacking the US capital, all he needed to say was, "stay there, stand by".

10.3.3 Structuring Roles through Interaction

Identity standards are interfaces. So, in taking on, or constructing, roles, we determine with what, whom, when, and how we might interact. Depending on our circumstances, we have some leeway in negotiating our identities through our interactions, which then mediate our relationships, not only to others but also to society as a whole (Hogg et al., 1995; Burke and Stets, 2009, 4; Tilly, 2016). When roles are instantiated in a social structure, they can be conceived of as a platform of interdependent, standardized positions (Merton, 1949; Katz and Kahn, 1966; Stryker and Burke, 1982). In any given case, the roles available to people is a function of the state and structure of society. In chaotic environments, for example, there are neither structures nor stable roles, as Alice learned to her dismay. On the other hand, in autocratic societies, such as China, Iran, and North Korea, standard roles are strictly prescribed. Mid-way between these two states of affairs is the case of Victorian society, where—as depicted so well in Jane Austin's novels—strictly prescribed manners kept deviant behavior in check (see Chapter 9, for a discussion).

Role congruence is a measure of the extent to which role expectations and role performances coincide. Congruence is most common in pre-modern societies, as well as in some rural communities today (Cote and Levine, 2002, 1; Collins, 2005). Within these settings, the pace of change is extremely slow, allowing for the incremental adaptation of roles to the environment. Additionally, roles are likely to be constrained, and strictly reinforced, by communal norms. As a result, children will generally assume the roles of their parents from whom they will also gain the knowledge and experience necessary to effectively enact their obligations (Fukuyama, 2018, 440).

In contrast to stable societies, identity formation is far more complicated in today's fast paced, complex world (Giddens, 1991; Luhmann, 2017). The number and variety of role identities has not only increased exponentially; as significantly, the boundaries between identities have also become blurred. At the same time, the social structures to enforce and integrate these roles have not always kept pace (Della Porta and Diani, 2020). Thus, it is difficult for individuals to connect to others, and to integrate themselves within society, in any meaningful way (Giddens, 1991; Cote and Levine, 2002).

10.4 Group Identity and Interaction Ritual Chains

In response to isolation, and the loss of a secure identity, many individuals are combining forces, and turning to others with whom they can share their interests and concerns and reinforce their status, position, and power (Fukuyama, 2018; Della Porta and Diani, 2020, 7, 13). This is not surprising insofar as ego identities are becoming increasingly fragile in our rapidly evolving, complex society. As Fukuyama (2018) has described:

> ... as important as material self interest is, human beings are motivated by other things as well, motives that better explain the disparate events of the present. This might be called the

politics of resentment …. In all cases a group, whether a great power such as Russia or China or voters in the United States or Britain, believes that it has an identity that is not being given adequate recognition—either by the outside world, in the case of a nation, or by other members of the same society. Those identities can be and are incredibly varied, based on nationality, religion, ethnicity, sexual orientation, or gender. They are all manifestations of a common phenomenon, that of identity politics. (18–19)

In contrast to the dyadic process of self-identification, identity formation within a group entails a series of interactions, negotiations, and performances that generate energy, commitment, and momentum to pursue the group's agenda. Building on the classic works of Emile Durkheim (1984) and Erving Goffman (1967), Randall Collins describes the process of group identification as the product of interaction ritual chains (Collins, 2005). According to Collins, in such groups, mutual focus of attention occurs, and emotional entrainment builds up, so that self-generating feedback processes give rise to moments of compelling emotional experience. These experiences serve as magnets of cultural significance where "culture is created, denigrated, or reinforced" (Collins, 2005, 12–13). Such rituals are transformative; they "take some emotions as ingredients and transfer them into other emotions", so as to fuel concerted group action (Collins, 2005, 13).

In contrast to highly organized protest groups, such as the Sierra Club and the ACLU, which are linked together by formal role relationships, today's movements are held together by standards that, taking the form of symbols, circulate and re-circulate through less formal, ritual interactions (Holland, 2012). Group symbols might include, for example, narratives, mottos, flags, tokens, logos, guns, dress codes, labels, etc. As the cycles of interactions go round and round, the group itself becomes a symbol of the enterprise and its purpose (Collins, 2005).

Protest groups not only bolster individuals' self-esteem by linking them to a group's identity. Group activity, and the newcomers energized by it, strengthen the group's image and legitimize its role. Unlike formal organizations, which have a well-defined structure, protest groups are typically found in loosely constructed networks, which are far more flexible (Collins, 2005; Tilly, 2016). Not being confined to a specific place, group networks can extend their reach and coordinate with distant, like-minded people to further enhance their prowess. One need to only witness the recent string of protests across American campuses on behalf of Palestine.

When organized into network platforms, groups also benefit from network effects (Shapiro and Varian, 1998; Rohlfs, 2001; Arthur, 2014). That is to say, the more people who join a group, the more others will be attracted to it. Given this advantage, groups greatly expand as they become more attractive to others. As competitors fall by the wayside, the dominant group gains influence as well. Understanding how identity-related grievances motivate such group processes, and set into motion disruptive activities, is vitally important today, as we increasingly experience what Fukuyama characterizes as grievance politics (Tilly, 2016; Fukuyama, 2018; Greenfeld, 2019).

10.5 The Emergence of National Identities

The emergence of national identities gave rise to complexities of far greater proportions, as can be seen, for example, throughout the Middle Ages (Bendix, 2017). During this period, identities were tied to the three estates—the nobles, churchman, and the commoners, as well as to the social hierarchies governing interactions within each category. Relationships within each hierarchy were based on status, with those at the top entailed exercising power over those on the rungs below. However, power holders were not without constraints. Ruling on behalf of, and in the name of, God, they were held to account, if only to a certain degree. On the other hand, commoners, while skirting nobles' dictates at the margin, freely submitted to their authority, given their dependence on the nobility for their livelihood and security, and their obligation to be obedient, as prescribed by the Church. As Bendix describes, it was these congruent role expectations—an implicit bargain of sorts—that, despite many isolated ups and downs, generated legitimacy for the governing system, and hence relative stability for the Middle Ages as a whole (Bendix, 2017).

As described below, the French Revolution and the emergence of nationalism and national identity not only rendered asunder these relationships and the standards that governed them; they also undermined the system's legitimacy. For, with the beheading of Louis XVI in 1793, sovereignty was transferred from the Estates General to the National Assembly which put into place The Declaration of the Rights of Man and the Citizens, abolishing the feudal system, as well the role relationships embedded in it (Blanning, 2002, 2007; Bell, 2003; Hibbert, 2012; Rowe, 2013; Doyle, 2018; Hastings, 2018; Greenfeld, 2019; Popkin, 2019). No longer mediated by, and tied to, established social structures, French citizens were on their own, no longer subject to, and guided by, social norms and collective responsibilities. The result, as described below, was a major phase transition, leading—with the collapse of Revolution—to a new, imperial form of government under Emperor Napoleon Bonaparte who, upon coming to power, instantiated a whole new set of standards legitimizing his regime.

10.6   National Identity and Revolution in 18th Century France

Troubled times and rapid social changes can severely undermine role identities with negative consequences to follow. Such was the case in the latter half of the 17th century in France, when a failed attempt to reorganize France's social structure led to the overthrow of the monarchy, and the onset of the French Revolution.

10.6.1   Legitimacy Undermined

In the latter half of the 17th century, the Middle-Age governance structure, sustained by congruent role expectations between rulers and commoners, began to crumble, along with the implicit bargain that shared norms entailed. Under considerable financial stress, the ruling classes could no longer meet the basic needs of the commoners, who—notwithstanding the greater demands placed upon them—felt less obliged to serve their superiors.

Exacerbating this state of affairs, life was worsening in the eyes of many. With the Nation's coffers verging on empty, financial problems were a major concern (Rapport, 2008, 624; Hastings, 2018, 22). Louis XIV had not only left France in great arrears; he had also decreased its borrowing capacity. Aggravating the situation was a loss of trust in the Monarchy due to discriminatory tax rates and several financial scandals and mishaps (Blanning, 2007, 557). As Hastings (2018) recounts:

> The politics of financial scandals became a virtual spectator sport in pre-revolutionary France, enhancing the destabilizing effect. Greedy land and currency speculators vied with corrupt government officials as the objects of popular gossip and scorn. (22)

Not surprisingly, the country suffered extensive poverty due to a population increase that coincided with poor harvests, leaving peasants living on the margin. Elites were, at best, indifferent to the peasants' plight, typically blaming peasants for their fate. Describing the situation, Doyle (2018) notes:

> Bands of roving vagabonds struck terror into the hearts of isolated farmers, and the streets of towns swarmed with beggars. The poor ... numbered at the best of times almost a third of the population: eight million people. In bad times two or three million more might join them, as crops failed, and jobs disappeared. (13)

Conflagrations related to religious roles exacerbated the country's tensions. At stake was the role of the Jansenists, a protestant-oriented Catholic order that encouraged ordinary people to ponder religious questions and contribute to religious discussions. To encourage a public dialogue, the Jansenists published a broadsheet, *Nouvelles Ecclesiasticus*, offering religious news and comment. These activities threatened the Pope, the Jesuits, and the Monarchy. Responding to the Jansenist efforts with force, the Pope issued the bull *Unigenitus,* labeling Jansenism a heresy and threatening to deprive its followers of the sacraments. However, given the Jansenists' large following, and the public's resentment of Papal interference in French church affairs, these efforts backfired (Farge, 2007, 43–53; Munck, 2005, 276; Rapport, 2008, 622–624; Doyle, 2018). For, Jansenists' losses in religious authority were compensated by gains in political influence, which they employed to engage the public on their behalf in the lead-up to the revolution (Blanning, 2002, 392; Blanning, 2007, 303–304; Farge, 2007;, 43–53; Rapport, 2008, 622–624; Doyle, 2018). Strengthening the Jansenist's position, their major rivals—the Jesuits, who had previously dominated French religious affairs, and were allied to the monarchy and the conservative nobility, were expelled from France in 1793–1794.

10.6.2 The Collapse of the Monarchy

From the 17th century until the middle of the 18th century, French kings had ruled as absolute monarchs, controlling the major decisions of their realms (Munck, 2005; Beloff, 2014). By the eve of the 18th century, however, the monarch's role had been abolished, and the country had descended from its height, as an absolutist power, to a state of Terror. Because politics and power had been organized in a tightly unified, hierarchical structure, the collapse of the monarchy took with it most of those who had been party to the Crown.

Surrounding himself with sycophantic courtiers at his court at Versailles, Louis XI—the Sun King—personified absolutist rule (Wilkinson, 2014, 18). Assuming power at age 23, he had demanded that all decisions, whether minor or major, be reviewed and approved by him, personally. To secure his autocratic position, Louis abnegated the role of the hereditary nobility in government, favoring instead professional nobles, who were dependent exclusively on him for their statuses and rewards (Popkin, 2019, 41). Explaining this choice to his son in his memoirs, Louis wrote:

> To be perfectly honest with you, it was not in my interest to choose persons of greater eminence. It was above all necessary to establish my own reputation and make the public realize, by the very rank of those who I selected, that it was not my intention to share my authority with them.
>
> *(Wilkinson, 2014, 86–87)*

To keep nobles and aristocrats loyal, while depriving them of their authority, Louis drew them to his Court at Versailles, where he could continuously monitor their behavior, and entice them with titles, money, and patronage instead of governmental offices (Greenfeld, 2019; Popkin, 2019). As importantly—much as Randall Collins has described with respect to the force of interaction rituals—Louis orchestrated and conducted the flow of life at Versailles. With everyone close together in physical proximity, the aristocracy's behavior could be constantly reinforced, not only through rigidly prescribed interactions but also through constant displays of luxury and wealth, which symbolized and signaled the Sun King's authority (Burke, 1992; Blanning, 2007; Wilkinson, 2014; Hastings, 2018). By so doing, Louis assured that his power stemmed not only from his blood lines as a Capet but also from his personal charisma, which he studiously exhibited in his elaborate performances at Versailles (Burke, 1992; Collins, 2022, 153). Describing Louis's rendition, Wilkinson, echoing Goffman, writes, "At court he was perpetually on show, a professional front man, presenting a dazzling man to the world" (Wilkinson, 2014, 100).

To maintain control, Louis's maintained strict protocols. Each person in his entourage had a distinct, prescribed role. For instance, Louis imposed rules and regulations of precedence, hierarchy, ritual, and formality, which were directly linked to power (Wilkinson, 2014, 101). As Wilkinson (2014) explains:

> Everything was ordered by protocol. Even in the royal chapel, in the presence of the God who died for all, the worshippers were graded according to rank, devoutly facing the king rather than the alter. In processions or at official receptions, precedence was sometimes only settled by duels. At meals people sat where they were entitled to sit or did not sit at all. And the various types of furniture were of profound significance: chairs with or without arms, stools, benches, and so forth. (102)

Louis XIV died in 1715. His successors Louis XV and Louis XVI had an identity problem. Not being as charismatic or energetic, they failed to meet their subject's role expectations. For, unlike Louis XIV, who had worked assiduously throughout his tenure, Louis XV and Louis XVI were considered passive, lazy, and dissolute. Upon Louis XIV's death, his offspring remained at Versailles, isolated from people and events. Hence, they lacked the throng of supporters that had catered to Louis XIV and signaled his supremacy. In fact, absent Louis XIV, the courtiers had little reason to stay at Versailles. Returning to Paris, they sought out new patronage and opportunities. (Doyle, 2018).

The Kings' insulation from Parisian life helps to explain Louis's successors' inability to appreciate—much less to adapt to—changes in French political culture (Rowe, 2013, 172; Popkin, 2019, 10). As Blanning (2007) has described:

> …not the least important reason for the collapse of the French monarchy was the repeated failures of Louis XV and Louis XVI to modernize their image and relegitimate their regime. Isolated from the public at Versailles, apparently wallowing in self indulgent luxury, they allowed the representational culture of Louis XIV to become dysfunctional, a dying weight that eventually dragged them down to perdition. (318)

10.6.3 Challenges to the Identity of Nobles and the Aristocracy

The French aristocracy, which had prospered under absolutism in the first half of the 17th century, suffered a considerable blow in its second half (Blanning, 2000, 81; Greenfeld, 2019). According to Greenfeld, the old order was suffering from status anxiety, as wealthy newcomers became more prevalent under the French system of veniality, which allowed them to buy their offices (Greenfeld, 2019, 44). Consider that up till the mid-18th century, nobles and aristocrats had executed complete authority and seigneurial rights in their own territories based on feudal laws. Ownership of land was, in fact, a crucial requirement for noble status. But social mobility, fostered through the widespread practice of veniality, chipped away at the nobility's wealth and traditional way of life. In fact, it was the resentment harbored by the traditional nobility towards the newly established nobles, who had earned their status not by virtue of their blood ties but as a result of their service to the state, that contributed not only to the revolt of the Frond in the mid-17th century, but also to Louis XIV's doubts about the aristocracy's loyalty (Greenfeld, 2019).

Adding to the aristocracy's concerns, Louis XIV—alarmed by the revolt of the Frond—had appointed government intendants, directly responsible to the monarchy, to positions in regional governments. This step was especially appalling to hereditary, noble families who had served as provincial governors (Wilkinson, 2014, 37–38). Under this new regime, many aristocrats succumbed to poverty. Those who, seeking to maintain their positions and supplement their income by venturing into business or trade, forfeited their nobility status as well as the benefits and privileges derived therefrom (Scott, 2008, 206–207). As Greenfeld (2019) explains:

> … the constant threat to the aristocratic status, undermined by the loss of political influence, the swelling of the ranks of the nobility, and the inflation of titles which now could be bought for money and made noble de race legally equal to low-born tradesmen who had accumulated enough of it, and the contemptuous attitude of the crown, had dire consequences for a society whose elite was affected in this manner. (44–45)

Looking for new identities and bases of support, several aristocrats began to reevaluate the Monarchy's role performance (Schama, 1990; Doyle, 2018; Greenfeld, 2019). They widely aired and intensely discussed their views among Enlightenment philosophers and other influentials in debates about the institutional contours of the French regime, and, in particular, the role of the Church (Munck, 2005, 373–374). A number of these aristocrats became active in the revolution (Schama, 1990; Doyle, 2018; Popkin, 2019).

10.6.4 The Decline of the Church

The Catholic Church was secure in its position throughout the first half of the 17th century, given the mutual interests of the church and state. In the aftermath of the Thirty Years War, Louis XIV believed that a united church was essential to assure peace in France. And, in fact, this was one reason why, in 1685, he revoked the Edict of Nantes, a declaration that had protected French Protestants for a century (Rapport, 2008, 613). Moreover, Louis XIV's closest advisors and ministers were drawn from the high clergy, while all bishops and archbishops were nobles. As well, next to the King, the clergy held the highest rung of the status ladder. However, while maintaining good relationships with the Papacy, Louis was nevertheless willing to support the French clergy in conflicts with the Pope.

Hence, the demise of the Catholic Church in France had less to do with politics, and more to do with two major social forces—the rise of nationalism and the Enlightenment. These events not only altered the hierarchical role relationships between the Church and Monarchy; they also reconfigured the ties between the Monarchy and the French people. Because proponents of nationalism called for the elimination of all status distinctions, it ruptured all hierarchical ties

(Greenfeld, 2019). At the same time, the Enlightenment provided the philosophical rational for dismissing the Church's role and reconstructing the government so as to guard against tyranny. Together, these developments severed the chain of ties that had linked the individual up through a hierarchy of status roles to the Monarch, providing thereby an opening for new political configurations (Munck, 2005, 277).

10.6.5 The Rise of the Citizen

The strains within the French body politic were reinforced and extended by the emergence of a public sphere, which surfaced along with the expansion of literacy and the growth of print media (Blanning, 2002, 193; Burke, 2005, 64; Farge, 2007; Rapport, 2008, 630; Schaich, 2008). By making political information and commentary widely available, the public sphere not only fostered political discussion; it also helped to transform individuals into engaged citizens. As Blanning (2000) describes:

> Situated between the private world of the family and the official world of the state, the public sphere was a forum in which previously isolated individuals could come together to exchange information, ideas and criticism. Whether communicating with one another over a long range by subscribing to the same periodicals or meeting face to face in a coffee house or in one of the new voluntary associations, such as a reading club or a Masonic Lodge, the public acquired a weight far greater than the sum of the individual members. It was from the public sphere that a new source of authority emerged to challenge the opinion-makers of the old regime: public opinion. (4)

Conversations within the public sphere were seditious over the long term, the fact of which the monarchy was aware (Farge, 2007; Schaich, 2008, 241; Popkin, 2019, 56). To monitor the situation, Louis XV set up a network of spies (*touches)* dispersed across Paris to document what people were saying. As Farge (2007) notes:

> … they flitted about, listening to exclamations and comments and transcribing them in reports which they sent regularly to the lieutenant general of police who then, in his weekly visit to the king, kept the monarch informed about the current climate. (3)

Nationalism also altered the third estate's notion of its role as citizens. As Hastings maintains, nationalism provided a form of group identity that "was rooted in a powerful sensation of belonging, a sense so compelling that, when fully articulated, it [overrode) all (or almost all) individual attachments and markers of identification" (Hastings, 2018, 2–3; see also Collins, 2020). Hence, if all French people were identical with respect to their nationality, then they must be equal in terms of their citizenship (Rowe, 2013, 172; Greenfeld, 2019). It was thus that nationalism paved the way for a revolutionary coalition of citizens—the *sans culottes,* who played a major role in moving the revolution to extremes. At the same time, nationalism allowed individuals to define their own identities without reference to the church and secular authorities. This was a momentous development. As Blanning (2007) contends:

> Now the issue was well and truly joined. In effect, what the third estate did during these great days in June was to proclaim the principle of national sovereignty and to claim for themselves the right to exercise it. This was the true beginning of the French Revolution. (321)

10.7 Failure to Achieve Collective Action

While it is one thing to stand together in opposition to a particular policy or position, it is quite a different matter to agree on what should follow in its stead (Collins, 2020). Such was the case in France after the Revolution (Palmer, 2005). As Doyle (2018) explains:

> Over the winter of 1788 and 1789 hardly anybody in France regretted the passing of the old political order. It had failed or let down too many people. Everyone assumed that change could only be for the better. But the process of working out what change there was to be took place in an atmosphere made tense by [the] acute and worsening economic crisis. (87)

Achieving collective action was, therefore, highly problematic.[1] While the French revolutionaries shared a commitment to a common goal—a democratic and egalitarian France—they differed not only on how to achieve it, but also on how, in the process, to maximize their own personal values and interests (Palmer, 2017). Thus, with the elimination of the monarchy's authority, the delegates turned their wrath against one another (Palmer, 2005; Tackett, 2015).

The first task of the revolutionaries was to create a constitution—all agreed to that. However, creating a governing constitution that was acceptable to all parties was an arduous process, given the wide range of issues and high stakes involved. Such an effort entailed, for example, not only creating new institutions but also new sets of roles and role relationships to govern how the occupants of such positions perform their tasks and interact with one another. The social structure, thereby created, defined the power relations between and among the actors involved, so personal ambitions complicated and inhibited a consensus (Palmer, 2005; Tackett, 2015).

Notwithstanding the difficulties entailed, members of the French Assembly committed themselves to this task on June 20, 1789, when they took the "Tennis Court Oath", requiring members of the National Assembly (the former Estates General) to remain together as a group, and to convene whenever necessary, to produce a new constitution (Hibbert, 2012, 49–51). It was a daunting, two-year ordeal, which required compromise and collective action.

10.8 The Politics of Chaos

To appreciate the problem of governance in post-revolutionary France, one need to only consider the numerous cleavages that characterized French society. Consider that France was divided into diverse geographic regions, each of which had distinct histories, customs, and languages. Rural and urban areas were also at odds, with Paris at the center playing a domineering role and causing resentment in the periphery, where calls for federalism resounded. The urban lower middle classes, which comprised the zealous *sans-culottes* as well as peasants in the countryside, were both suspicious and resentful of those with status and wealth. Even though the revolution had decreed equality for all, the dreadful memory of feudalism lurked in the shadows, so that every provocation and rumor was met with violent uprisings. While a majority of the decision makers in the National Assembly (subsequently the Constitutional Convention) shared legal and professional backgrounds as well as a devotion to the heady doctrines of nationalism and the Enlightenment, they adamantly disagreed with one another about the future social order, especially the fate of the Church, the role of Monarchy, the allocation of property rights, the place for federalism, and the decision to go to war.

Two major factions coalesced around these diverse positions: the radical Montagnards, who favored eliminating the King and aligned themselves with the *sans culottes*,[2] and the more moderate Girondists, who favored the support of property rights and a constitutional monarchy, while disdaining the common people. Thus, instead of finding a common way forward, these two factions erupted into a fierce, and at times, vicious rivalry, with each side seeking to rid the other from the political arena (Hilbert, 1981, 182–183; Palmer, 2005; Tackett, 2015; Doyle, 2018; Popkin, 2019).

Retaliating against attacks by the Girondists, who had initially enjoyed a majority in the Convention, the Jacobins/Montagnards, which included Danton, Marat, and Robespierre, employed their exceptional oratorical and rhetorical skills not only to co-opt the middle-of-the-road convention delegates (The Plain) but also to incite the *sans-culottes* to carry out violent protests on their behalf. To curry favor, and maintain the *sans-culottes'* trust, the Montagnards replicated many of the common peoples' styles and aspects of behavior. As Tackett (2015) notes:

There was … evidence of a subtle evolution in the culture of the deputies, expressive of the new democratic ideals and—at least for some representatives—of a self conscious identification with the popular classes. Many had now taken to changing their mode of clothing…. Over the months since 1789, powdered wigs, knee breeches, silver buckled shoes and elegant vests had all been abandoned by many of the Revolutionaries, to be replaced by long leather boots and bought coats they might have worn while hunting or tromping about. (259)

The Montagnards strategy of triangulation proved highly successful. The violence in the streets, and the deaths and imprisonments that ensued, convinced more moderate delegates to make concessions to the Montagnards and to purge recalcitrants from the Convention. It was in this way that the Montagnards were able to eliminate the Girondists and other moderate delegates in the turbulent period after the King's execution, thereby moving the Convention further and further to the left. This pattern was repeated throughout the course of the revolution so that, as enemies were eliminated, the seat of power shifted from the Convention to the more restricted Paris Commune and the Committee of Public Safety, both of which the Montagnards controlled.

One important consequence of this repositioning was "government by Terror". Once France had declared war on much of Europe, the country was suddenly faced with opposition not only from within but also from without (Tackett, 2015). Obsessed by the prospect of foreign intervention, and believing spies to be everywhere, Robespierre declared it necessary to put the constitution on hold, and to rule by Terror for the duration of the war (Tackett, 2015; Palmer, 2017).

To execute the Terror, government power was shifted to the Committee on Public Safety, which reported to the Constitutional Convention. Here, policy shifts were laid out by Robespierre and others *via* elaborate speeches laden with fierce, revolutionary rhetoric. Opposition, which was considered anti-revolutionary, was dangerous, as those who spoke out were likely to be purged from the convention, jailed, and/or subsequently executed.

In the provinces, the Terror was carried to extremes. Members of the Convention and the Committee for Public Safety were sent on missions to the provinces where they oversaw the arrest, trials, imprisonment, and execution of suspected counter revolutionaries. With such a range of responsibilities, missionaries took things into their own hands, often discounting and disregarding orders from Paris (Palmer, 2017). Thus, although Robespierre had explicitly instructed those on mission to refrain from persecuting Christians, extremists, such as Collot d'Herbois, Lacoste, and Baudot, did so vehemently. While those in Paris may have been alarmed by such flagrant violence, they could do little about it. As Palmer (2017) explains:

> Presented with an accomplished fact, [the Committee] had to give its approval. No one in authority could bear the stigma of moderation. No one could seem to befriend enemies of the people. No one claiming to be true to the Revolution could safely be less advanced than another who made the same claim. (172)

And so, the Terror went on unleased until 17,000 citizens had been executed. One must wonder, then, what crimes were so many people guilty of? Ironically, most of those imprisoned and executed were guilty not for what they had done or said but rather for *who they were—that is to say, their identities.* Because these prisoners were prevented from defending themselves, it was their accusers who defined their identities. To rationalize their behavior, they stereotyped their opponents with incendiary labels, characterizing them as profiteers, hoarders, counter revolutionaries, traitors, and foreign conspirators without providing evidence for their claims. Nor did they stipulate standards delineating what anti-revolutionary behavior entailed. Lacking standard criteria, innocent republican citizens could not vindicate themselves.

The Terror was not confined to the countryside. Within the Convention, delegates accused each other to avoid being targeted themselves. Even Danton, a charismatic and poplar patriot, as well as a longtime colleague of Robespierre, could not avoid the guillotine. Delegates, wearied of the mass killings and worried for their own safety, began to question Robespierre's leadership.

In the summer of 1894, Robespierre returned, after a three-week absence, to the convention floor where he delivered a rambling speech, in which he declared the need to, yet again, cull the delegates without, however, naming those at risk. In fear for their lives, the delegates challenged him, and—after a major skirmish—Robespierre was apprehended.

10.9 Lacking Standards and a Stable Social Structure

How did French patriots metamorphize from being ardent followers of the Enlightenment to becoming the executioners of the Terror? There are many partial explanations: conflicts and bitter rivalries among the patriots, the betrayal of the King and other heroes such as Mirabeau and Lafayette, disillusionment with Catholicism, the violence of the *sans-culottes,* shortages of food, fear of the Terror, and the foreign plots to revanche the monarchy. But all these matters would have been less incendiary, had there been common norms and standards to which the Patriots might have grounded themselves.

To be sure, developing standards in a revolutionary milieu is a futile exercise, especially when, as in 18th century France, all roles and status distinctions had been dissolved. The Patriots had to establish and maintain their own role designations through fierce competition and resorts to violence, a process made all the more difficult given the country's overlapping administrative apparatus and the existential threat of the *sans-culottes.* Making matter worse after the King's execution was the fact that loyalty to the Revolution was the only unifying principle holding the Patriots together. And so, the revolution had to go on in a climate of intense fear and distrust. As Randall Collins (2005) has pointed out, the emotional ardor and drama that accompanied the revolution entrained the participants such that they could not retreat.

Notes

1 As described in Chapter 2, collective action will most likely occur when there is a shared vision, a commitment to a common goal, congruent values and interests, consistent standards of behavior and interaction, as well as positive feedback.
2 *Sans culottes*.

References

Arthur, Brian (2014) *Complexity and the Economy*, New York, NY: Oxford University Press.
Bell, David A. (2003) *The Cult of the Nation in France—Inventing Nationalism 1680–1800*, Cambridge, MA: Harvard University Press.
Beloff, Max (2014) *The Age of Absolutism, 1660–1815*, New York, NY: Routledge.
Bendix, Reinhard (2017) *Nation Building and Citizenship, Studies of Our Changing Social Order*, New York, NY: Routledge.
Biddle, B.J. (1986) "Recent Developments in Role Theory," *Annual Review of Sociology* 12, 67–92.
Blanning, T.C.W., ed. (2000) *Eighteenth Century Europe*, London: Oxford University Press.
Blanning, T.C.W. (2002) *The Culture of Power and the Power of Culture*, Oxford: Oxford University Press.
Blanning, T.C.W. (2007) *The Pursuit of Glory*, New York, NY: Penguin.
Blumer, Herbert (1986) *Symbolic Interactionism: Perspectives and Method*, Oakland, CA: University of California Press.
Bowker, Geoffrey C., and Susan Leigh Star (2000) *Sorting Things Out: Classification and It Consequences*, Cambridge, MA: MIT Press.
Buchanan, Mark (2007) *Why the Rich Get Richer, Cheaters Get Caught, and Your Neighbor Usually Looks Like You*, New York, NY: Bloomsbury.
Burke, Peter (1992) *The Fabrication of Louis XVI*, New Haven, CT: Yale University Press.
Burke, Peter J., Timothy J. Owens, Richard E. Serpe, and Peggy A. Thoits (2003) *Advances in Identity Theory and Research*, New York, NY: Plenum Publishers.
Burke, Peter, and Jan Stets (2009) *Identity Theory*, New York, NY: Oxford University Press.
Carroll, Lewis (2009) *Alice's Adventures in Wonderland and through the Looking Glass*, Oxford: Oxford University Press.
Callero, Peter L. (2003) "The Political Self—Identity Resources for Radical Democracy," in Burke, Peter J. et al., eds. *Advances in Identity Theory and Research*, New York, NY: Prenum Publishers/Springer.
Carson, Lionel (2015) *Life in Ancient Rome*, New Word City, www.NewWordcity.com
Collins, Randall (2005) *Interaction Rituals Chains*, Princeton, NJ: Princeton University Press.
Collins, Randall (2020) *Charisma: Microsociology of Power and Influence*, New York, NY: Routledge.
Collins, Randall (2022) *Explosive Conflict, Time-Dynamics of Violence*, New York, NY: Routledge.
Cote, and Levine (2002) *Identity, Formation, Agency, and Culture: A Social Psychological Synthesis*, Yahweh, NJ: Lawrence Erlbaum Associates, Inc.
Davidson, Ian (2016) *The French Revolution*, New York, NY: Pegasus Books.
Della Porta, Donnatella, and Mario Diani (2020) *Social Movements, An Introduction*, New York, NY: John Wiley and Sons.
Doyle, William (2018) *The Oxford History of the French Revolution*, New York, Oxford University Press.
Dunstan, William E. (2011) *Ancient Rome*, Lanham, MD: Rowan and Littlefield Publishers, Inc.
Dupre, Louis (2004) *The Enlightenment and the Intellectual Foundations of Modern Culture*, New Haven, CT: Yale University Press.
Durkheim, Emile (1984) *The Division of Labor in Society*, New York, NY: The Free Press.
Erikson, Erik (1968) *Identity, Youth, and Crisis*, New York, NY: W.W. Norton & Company.
Farge, Arlette (2007) *Subversive Words: Public Opinion in Eighteenth Century France*, University Park, PA: Penn State University Press.
Faulkner, Neil (2013) *Rome, Empire of the Eagles*, New York, NY: Routledge.
Fukuyama, Francis (2018) *Identity, the Demand for Dignity and the Politics of Resentment*, New York, NY: Farrar, Straus and Giroux.
Giddens, Anthony (1991) *Modernity and Self Identity: Self and Society in the Late Modern Age*, Cambridge: Polity Press.

Goffman, Erving (1959) *The Presentation of Self in Everyday Life*, Edinburgh: University of Edinburgh Social Sciences Research Center.

Goffman, Erving (1967) *Interaction Rituals*, Piscataway, NJ: Aldine Transactions.

Greenfeld, Liah (2019) *Nationalism, a Short History*, Washington, DC: Brookings Institute.

Grewal, David Singh (2008) *Network Power, The Social Dynamics of Globalization*, New Haven, CT: Yale University Press.

Hastings, Derek (2018) *Nationalism in Modern Europe: Politics, Identity and Belong after the French Revolution*, London: Bloomsbury Academic.

Heather, Peter (2006) *The Fall of the Roman Empire, A New History*, New York, NY: Oxford University Press.

Hibbert, Christophe (2012) *The Days of the French Revolution*, New York, NY: Harper Collins Publishers, Inc.

Hogg, Michael A., et al. (1995) "A Tale of Two Theories: Comparison of Identity Theory with Social Identity Theory," *Social Psychology Quarterly* 58(4), 255–269.

Holland, John (2012) *Signals and Boundaries*, Cambridge, MA: MIT Press.

Johnson, Alison (2013) *Louis XVI and the French Revolution*, Jefferson, NC: McFarland and Co., Inc.

Kamm, Antony, and Abigail Graham (2015) *The Romans: An Introduction*, 3rd edition, edition. New York, NY: Routledge.

Katz, Daniel, and Robert L. Kahn (1966) *The Sociology of Organizations*, New York, NY: John Wiley & Sons.

Lidz, Victor (2021) "Afterword: A Functional Analysis of the Crisis in American Society, 2020," *The American Sociologist* 53, 241–242.

Luhmann, Niklas (2017) *Trust and Power*, London: Polity Press.

Mattingly, David J. (2011) *Imperialism, Power, and Identity, Experiencing the Roman Empire*, Princeton, NJ: Princeton University Press.

Mead, George H. (1967), *Mind, Self and Society: From the Standpoint of a Social Behavioralist*, edited and with an introduction by C.W. Morris, Chicago, IL: University of Chicago Press.

Merton, Robert (1949) *Social Theory and Social Structure*, New York, NY: Free Press.

Munck, Thomas (2005) *Seventeenth-Century Europe: State, Conflict and the Social Order in Europe, 1598-1700*, New York, NY: Palgrave.

Palmer, R.R. (2017) *Twelve Who Ruled*, Princeton, NJ: Princeton University Press.

Potter, David S. (2004) *The Roman Empire at Bay*, New York, NY: Routledge.

Rapport, Michael (2008), "France," in Wilson, Peter, H.A. *Companion to Eighteen Century Europe*, New York, NY, Blackwell Publishing.

Rohlfs, Jeffrey H. (2001) *Bandwagon Effects in High Technology Industries*, Cambridge, MA: MIT Press.

Rowe, Michael (2013) "The French Revolution, Napoleon, Nationalism in Europe, Chapter Seven," in Breuilly, John, ed. *The Oxford Handbook: The History of Nationalism*, Oxford: University of Oxford Press.

Schaich, Michael (2008) "The Public Sphere," Chapter Eight, in Wilson, Peter E., ed. *A Companion to 18th Century Europe*, New York, NY: Blackwell Publishing.

Schama, Simon (1990) *Citizens: A Chronicle of the French Revolution*, New York, NY: Vintage.

Sen, Amarta (2006) *Identity and Violence*, New York, NY: W.W. Norton and Company.

Serpe, Richard T., and Sheldon Stryker (2011) "The Symbolic Interactionist Perspective and Identity Theory," in Schwartz, Seth, et al. eds. *Handbook of Identity Theory and Research*, New York, NY: Springer.

Shapiro, Carl and Hal Varian (1998) *Information Rules, A Guide to the Network Economy*, Cambridge, MA: Harvard Business Review Press.

Southern, Patricia (2012) *Ancient Rome, the Rise and Fall of an Empire*, Gloucestershire: Amberley Publishing.

Spears, Russell (2011) "Group Identity, the Social Identity Perspective," in Schwartz, Seth, et al. eds. *Handbook of Identity Theory and Research*, Berlin: Springer.

Stets, and Burke (2000) "Identity Theory and Social Identity Theory," *Social Psychology Quarterly* 65, 224–337.

Stryker, Sheldon, and Peter J. Burke (1982) "The Past, Present, and Future of Identity Theory," *Social Psychology Quarterly* 63(4), 281–297.

Tatlow, Didi Kristen (2013) "Letter from China, A Moniker Only a Mister Could Like," *New York Times*, August 27, 2013, p. A7.

Tackett, Timothy (2015) *The Coming of the Terror in the French Revolution*, Cambridge, MA: The Belknap Press of Harvard University Press.

Theist, Peggy (2003) "Personal Agency in the Accumulation of Multiple Role Identities," Chapter 12, in Burke, Peter, et al. eds. *Advances in Identity Theory and Research*, New York, NY: Plenum Publishers.

Tilly, Charles (2016) *Identities, Boundaries and Social Ties*, New York, NY: Routledge.

Weber, Maximillian (1947) "The Nature of Charismatic Authority and Its Routinization," in *Theory of Social and Economic Organization*, translated by A.R. Anderson and Talcott Parsons, Washington, DC: Free Press.

Wilkinson, Richard (2014) *Louis XIV*, New York, NY: Routledge.

11

THE ARTIST AS STANDARDS BEARER

11.1 Artistic Agency and the Role of the Standard's Bearer

No man is an island, not even the "lone" artist. Despite the deep-seated presumption cultivated so well in the Romantic Era that artists operate in their own worlds inspired by their inner muses, recent scholarly works written in the socio-historical tradition have clearly demonstrated that artists—defined here broadly—are very much embedded in, and inspired by, their material surroundings and the specific contexts in which they find themselves (Becker, 1982; Gardner, 1991; Bourdieu, 1993; Czikszentmihalyi, 1999; Sawyer, 2006; Hauser, 2011). That said, however, one cannot deny that artists have agency in their own right. In fact, artists play an important mediating role by producing artifacts that focus a society's attention on meaningful, shared symbols that allow for coordinated behavior and the emergence of a cultural identity.

In this sense, artists can be conceived as standards bearers. A standard bearer is an outstanding leader who carries a flag or emblem that coordinates and rallies followers around a common cause. The standard has typically been used and honored as a visual symbol of a state, military unit, or domination (Wikipedia; Cambridge Dictionaries Online). As such, standards have often been associated with the sacred.

The Romans made especially good use of standards as an appurtenance of war. In each Roman legion, the standard bearer carried an aquila, or eagle, used to direct troop movements in the field. Standards were so important, Romans took great care to protect them, going to extreme lengths to recover standards lost in battle. Thus, after the people of Germania defeated the Romans at the Battle of the Teutoburg Forest, the Romans spent decades trying to recover three lost standards taken by the enemy.

Just as standards bearers rallied troops, and coordinated their actions on the battlefield, so artists have helped to sway the turn of events by reinforcing or undermining social integration through cultural memes. As sociologist Daniel Bell described, culture is a system of symbols, beliefs, behaviors, and institutions that define and reflect the social reality of a community (Bell, 1976, 12). These cultural standards provide a consistent moral and aesthetic frame of reference. In this context, artists' works serve as a form of rhetoric that, more often than not, supports the powers that be by reconciling individual identities with their cultures (Woods, 2012, 30–31).

DOI: 10.4324/9781032721125-15

However, in periods of major upheaval, artists play a leading role in redefining and refueling the cultural milieu and, with it, social parameters for the future (Boime, 2014).

11.2 Pierre Bourdieu's Contested Art Field

For a theoretical perspective about how artists operate in their fields and establish agency through their works of art, consider the works of French sociologist Pierre Bourdieu. In *The Field of Cultural Production* (1993), Bourdieu lays out a theory that, while assuming the importance of social structure in constraining choices, still provides an explanation of how artists engage in generative action (Jenkins, 1992, 34; Berlin, 1999; Boime, 2014). According to Bourdieu, the relationship between artists and their environments is a dialectical one. During their socialization, artists are inculcated with certain tastes, dispositions, and worldviews. These predilections, which Bourdieu characterizes as the *habitus*, are like a second sense. Creating their artwork through the lens of their *habitus*, artists help to reproduce and reinforce the existing order of things (Berlin, 1999, 32–33; Harris, 2011; Boime, 2014). Nevertheless, artists and their works are not isolated. Artists live and carry out their work within the context of an organized "field". Here, consensus regarding objectives does not preclude conflict. For fields are highly contested arenas in which actors compete not only for resources but also for legitimacy.

11.2.1 Artistic Fields

In opposition to the notion that artists exhibit special gifts setting them apart from others, Bourdieu claims artists develop their skills and ideas in conjunction with their participation in, and contestation of, their field of interactions. Actors engage, for example, with other artists, patrons, publishers, art critics, museums, journals, students, etc. (Bourdieu, 1993; see also Becker, 1982; Czikszentmihalyi, 1999; Sawyer, 2006). Generally speaking, art fields are self-contained and independent within their own spheres. However, when fields become a battleground of contesting parties, their autonomy is greatly reduced. Contending parties will seek to reinforce their positions by linking them to the power structures and deep-seated values and norms of society as a whole (Bourdieu and Wacquant, 1992). It is at this point that those in power have a unique opportunity to revisit and reevaluate performance in the art field[1].

11.2.2 Artistic Agency

Artists, however, are not without agency (Kristeller, 1990, xiii; Woods, 2012, 102). Although the cultural field is structured hierarchically in accordance with existing power relationships, its participants jockey for power and prestige. Disruptions restructure the field, providing openings for artistic innovations and social change. Artists' works legitimate one side of the power conflict or another. Artistic breakthroughs occur in conjunction with phase transitions,[2] bringing about fundamental changes throughout society (Harris, 2011; Hauser, 2011; Weiner, 2016).

11.2.3 The Art Field as a Complex System

Bourdieu's fields are complex adaptive systems (Holland, 2014). They evolve based on the interplay among their diverse parts, which are governed by a set of standards that are either reinforced or contested during those interactions. Much like niches, art fields have semi-porous boundaries, allowing them to interact with fields both at their same level (i.e., music, theater, literature) or at higher levels in the hierarchy of power (Holland, 1999). As artists adapt, they generate new standards reconfiguring the fitness landscape in the process. As in all complex adaptive systems, the dialectic between artistic fields and fields of power are contingent on specific histories and events (Scott and Simms, 2007, 10). To characterize the field's evolution, while capturing these contingencies, I depict artistic standards in three historical contexts: the Renaissance in Florence Italy, the Romantic era in Europe, and the art world of Weimar Germany. In each case, major artistic breakthroughs occur in the context of major social upheavals.

11.3   Emergent Creativity in Troubled Florence

Between the 13th and the 16th centuries, Florence was subject to major upheavals, ranging from foreign invasions, papal conflicts, peasant revolts, and the plague (Franceschi, 2008; Najemy, 2008b; Hauser, 2011, 88). Nevertheless, Florence was a font of unprecedented artistic achievements and revolutionary innovations (Joost-Gaugier, 2013, 17). These advances not only mirrored changes in political fortunes; they also served to legitimate them (Woods, 2012, 6). In a city priding itself for its republican virtues, yet ridden by intense internal conflicts, literature and the visual arts provided a cultural veneer that belied the internal power struggles and reinforced the city's humanistic reputation and self-identity (Najemy, 2008a, 3).

11.3.1   The Players and Stakes Involved

To appreciate how artistic rhetoric fostered a unified Florence, consider the players and stakes involved. In the early 13th century, Florence was ruled by aristocratic families from the countryside living in enclosed family compounds located in separate quarters of the city. These compounds consisted of heavily defended castles, fully equipped defense towers, family member residencies, servants' quarters, and space for a military contingent. Elites competed for dominance using every means at their disposal—bribery, kinship ties, patronage, network connections, marriage alliances, and, when necessary, assassination, and armed force (Kent, 2008; Najemy, 2008a). Conflicts were exacerbated by foreign intervention allowing dissenters to ally with foreign powers eager to gain a foothold in Tuscany. Internal peace was maintained through a balance of power, as overly aggressive families were subject to banishment or execution.

The Black Death greatly reconfigured Florence's social structure. Reaching its peak between 1340 and 1350, it took the lives of one half to two-thirds of the population, rich and poor alike (Franceschi, 2008, 107–110; Goldtwaite, 2009, 39). Immigrants from the countryside filled the population vacuum. Comprised of shopkeepers, retailers, teachers, lawyers, notaries, and artisans, they constituted a burgeoning middle class (Najemy, 2008a). Industrious and entrepreneurial, they built the textile and banking industries, making Florence a major economic power not only in Italy but also in Europe and the Levant (Najemy, 2008a, 6, 23, 113; Goldtwaite, 2009, 11–12).

11.3.2   Emerging Social and Political Divisions

With this new wealth came new social divisions and new aspirants for power (Najemy, 2008b, xxiv; Goldtwaite, 2009, 28). The *Popolo*, a heterogeneous group that included successful non-aristocratic merchants, shopkeepers, notaries, and artisans, challenged the aristocracy (Najemy,

2008a; Zorzi, 2008). Leaders were well off and highly educated. What differentiated them from traditional elites, and what held them together despite their diverse interests, was not only their pride in Florence but also a deep commitment to the philosophy and practice of humanism (Kristeller, 1990).

11.3.3 Revolution and Reaction

By the late 13th century, the *Popolo* had reached the height of its success. Organized into self-governing guilds based on professional practices, the *Popolo* functioned as an alternative government that excluded the noble class (Hibbert, 1974, 24). Participants were selected by lot to serve for two-month periods in the governing Signoria (Hibbert, 1974; Zorzi, 2008). Not surprisingly, traditional elites continued to seek a foothold in government. To this end, they manipulated the lottery system in their favor. At the same time, to gain legitimacy, they embraced—or at least appeared to embrace—humanism and the civic way of life (Najemy, 2008b, xxxi). Nowhere was this behavioral shift more evident than in their patronage of humanist texts and new art forms.

11.3.4 Humanism, an Integrating Force

The Florentines' fixation with humanism reflected their belief that they were heirs to Rome and defenders of republican liberty (Quillen, 2008, 28). To fulfill this historic mission, they sought out humanist writings of the Greeks and Romans, copying, translating them, and developing new scholarly writings and humanist learning materials[3] (Kristeller, 1990, 3). These literary texts were intended to provide the fundamental principles, or might I say standards, for developing and educating a literate citizenry capable of engaging in virtuous actions.

In the visual arts, Florentine humanism also played a normative role. Believing that Florence had set the standard for good governance, Florentines sought to demonstrate the city's primacy by investing in, and displaying artistic works, especially innovative building, and architecture projects (Goldtwaite, 1982). As patronage shifted from the Church to public and private funders, the city of Florence was totally transformed. According to Hauser, it was this drive to excel that accounts for Florence becoming "the only place in Italy where any important progressive activity [was] taking place in the world of art" (Hauser, 2011, 90–92).

11.3.5 The Brunelleschi Circle

Most notable among Florence's pioneering artists were members of the Brunelleschi Circle, whose work was grounded in humanism. Included in this group were Masaccio, Alberti, Donatello, and Ghiberti (Hutter, 2015, 59–60). In contrast to the metaphysical symbolism of the medieval period, these artists portrayed a more naturalistic understanding and representation of the empirical world, drawing on mathematics and science to develop a new way of representing space through linear perspective (Hauser, 2011, 2, 65; Woods, 2012, 333; Joost-Gaugier, 2013; Kleineler, 2015). Unlike the Gothic style, which organized space on a vertical basis, linear perspective required artists to locate their subjects horizontally and place them within a realistic architecture that brought the viewer's eye to the center of a painting, thereby providing a unity to the whole (Hauser, 2011; Joost-Gaugier, 2013, 71).

Alberti, a well-known humanist artist, writer, and philosopher, standardized this technique in his treatise, *Della Pittura* (1436), a work that was highly influential both within Italy and beyond.

11.3.6 Coopting Humanism

Attending to art and humanist literature, the nobility reinvented itself (Najemy, 2008b, xxxi). Instead of vying for power through military might, elites competed based on displays of their buildings and works of art, as well as their prowess in humanist rhetoric and learning (Goldtwaite, 2009; Woods, 2012). Much as Bourdieu explained with respect to dialectical turns of events, these wealthy patrons not only embraced new art and literature; they also adapted it to their own purposes (Najemy, 2008b, xxxii). Through their close ties to Renaissance artists and literati, the Florentine elite generated effective propaganda that, building on widely accepted cultural and religious tropes, helped disguise their power. Describing the impact of art on Renaissance politics, Muir notes: "the arts and ceremonies in Renaissance Italy *helped to create, not just record, the dominion of power*" (Muir, 2008, 228, emphasis mine).

11.4 Neoclassicism

Neoclassical art emerged during the Enlightenment (Honour, 1979; Brown, 2001, 9). Opposed to the ornamental, decorative style of Rocco art that was popular throughout the early 18th century, neoclassical artists sought to revive the seriousness of art by replicating the themes, narratives, and standards of antiquity. To create a sense of constancy and serenity, artists employed an austere linear approach aimed to recreate the heroic, covetable scenes of the ancient past (Walsh, 2017, 3). To avoid distractions, color was downplayed, as were rough and uneven brush strokes (Barker, 2015, 378; Walsh, 2017, 39). The goal was to achieve a standard of beauty that could be derived theoretically and thus taught and passed down to future generations (Crow, 1985; Boime, 1990; Blanning, 2012, 12; Barker 2015, 23).

11.4.1 Linking Politics and Art

These artistic standards were reinforced via a complex set of networks and institutions that functioned as a system of patronage (Barker, 2015, 25). This network fostered cultural themes and artifacts that served as propaganda for the ruling classes (Boime, 1990, 2014; Barker, 2015, 3). Louis XIV, for example, excelled at employing art to bolster his image (Blanning, 2010). At Versailles, every means of display aimed to create an aura of sumptuousness to buoy his persona and legitimise his status. Subsequent rulers and aristocrats continued this practice, providing the mainstay for artists to carry on their work (Barker, 2015, 6). Emperor Napoleon, for example, commissioned artists such as Jacques-Louis David and Antoine-Jean Gros to portray his exploits on the battlefield. Similarly, during the stressful post-Napoleonic period, Louis XVI continued to use neoclassic standards to signal calm and stability throughout his tenuous realm.

The neoclassical narrative was perpetuated by the linkages and the fluid boundaries between the artists and the ruling classes. The major venue for these linkages was the national academies of art, where budding artists congregated for instruction (Crow, 1985). By the turn of the century, there were more than one hundred such academies throughout Europe (Blanning, 2010, 13).

11.4.2   Setting the Standards

From the time of Louis XIV, the French Academy of Painting and Sculpture established art standards. The Academy was comprised of aristocrats who shared a common ruling-class perspective. They asserted that, because of their classical education, only they could represent the "public" interest (Crow, 1985, 3; Prickett, 2014, xxxviii). To promote their conservative aims, they developed a hierarchy of painting that ranked large-scale history paintings, which required classical training, at the top. Because history paintings were expensive to produce, they not only replicated the beauty and virtue of antiquity; they also reinforced the social hierarchy by restricting the participation of lower-class artists (Crow, 1985). As Barker (2015) describes:

> … "the antique" (as it was known) functioned as a cultural norm, setting the standards of good taste that distinguished the social elite from the working poor, who had not access to the body of ideas, texts and objects that constituted the classical tradition. Images and artifacts that lay outside that tradition were not considered art by these standards …. (23)

11.4.3   The National Academy

Membership and rank in the Academy depended on an artist's ability to demonstrate his prowess in bi-annual Paris Salon exhibitions as well as in the Grand Tour to Rome where the French had an affiliated academy and an active art community. The Academy exerted its influence by determining which paintings would be exhibited at the semi-annual Paris Salons; which consignments would be made and to whom; as well as which artists would be sent to Rome to compete for the Grand Prize. Under these circumstances, French artists were beholden to their patrons, so they typically catered to upper class neoclassical tastes (Crow, 1985, 2).

11.4.4   The Commercialization of Art

As markets came into play, patronage ties were loosened and artistic standards were increasingly set by private collectors, art dealers, and an extended, bourgeois "public" that kept abreast of cultural events and frequented the salons (Crow, 1985; Habermas, 1991; Blanning, 2012, 37; Barker, 2015, 26). These newcomers gained access to the art world through the media and networks, which were external to the Academy. These included newspapers, periodicals, coffee houses, art exhibitions, concerts, literary societies, reading clubs, museums, and so on (Salmi, 2008, 55–56; Blanning, 2012, 37; Barker, 2015). Especially influential were professional art critics who attended salons and published their evaluations of artists' works (Crow, 1985; Barker, 2015, 317). This expanded public not only served to provide a new base of income; it also paved the way for new artistic standards (Blanning, 2012, 37).

With the development of a commercial market, the debate over standards centered on who could best serve the public interest. As Crow (1985) has described:

> … the role of the new public space in the history of eighteen century French painting [was] bound up with a struggle over representation, over language and symbols and who had the right to use them. The issue was never whether that problematic entity, the public, should be consulted in artistic matters, but who could be legitimately included in it, who spoke for its interests, and which or how many of the contending directions in artistic practice could claim its support. (50)

Historical factors intensified the conflict, so that what might have been esoteric questions of style and subject matter were caught up in a power struggle (Crow, 1985, 50). Participants recognized they were living through a "watershed in history" (Berlin, 1999, 30–31; Salmi, 2008; Blanning, 2012, x). While some despaired, others saw an opening for a new Utopia (Lowy and Sayre, 2001; Besier, 2004). As French author Alfred de Musset (2008) described in his *Confessions of a Child of the Century* (1836):

> ... behind them a past forever destroyed, still quivering on its ruins with all the fossils of centuries of absolutism; before them the aurora of an immense horizon, the first gleam of the future; and between these two worlds—like the ocean which separates the Old World from the New—something vague and floating, a troubled sea filled with wreckage, traversed from time to time by some distant sail or some ship trailing thick clouds of smoke; the present, in a word, which separates the past from the future, which is neither the one nor the other, which resembles both, and where one cannot know whether, at each step, one treads on living matter or on dead refuse. (12)

Just as the political and economic worlds were upended, so too was the cultural landscape. This rupture gave rise not only to a shift in artistic standards but also to a corresponding change in the structure of the art world itself (Barker, 2015, 517). As Blanning (2012) notes:

> In just two or three generations, the rule book of the classical past was torn up. In its place came *not* another set of rules, but a radically different approach to artistic creation that has provided the aesthetic axioms of the modern world, even if a definition of romanticism has proved elusive. (x0)

Much as Bourdieu might contend, this breach was a contestation of the art "field". At stake was the neoclassical tradition vs. the Romantic vision (Brown, 2001, 4).

11.5 Romanticism Emerges

Romanticism emerged within this changed fitness landscape driven by the commercialization of art as well as by a growing, negative reaction to the neoclassical tradition. Instead of viewing art as being based on standardized, codified visual symbols, Romantic symbols were individualistic, and often based on an artist's own personal experience (Honour, 1979, 17). Feeling themselves hemmed in by neoclassic norms, artists—in the late 17th early 18th centuries—turned their sights inward, seeking inspiration and truth within their own, individual souls (Berlin, 1999). As Honour (1979) explains:

> For now, the work of art—painting, poem, or musical composition–came to be regarded not simply as a reflection of reality or the embodiment of an immutable and rationally conceived ideal, but as an insight into the life of things and, perhaps, a means of lightening the 'burthen of the mystery ... the heavy and the weary weight of all this unintelligible world'. (21)

11.5.1 Individualistic Styles

Unlike Academy practices, whereby artists' works were classified according to a rigid, hierarchical scale of value, Romantics considered all modes of art equally worthy. Refuting neoclassical rigid practices, they rejected the standards of beauty as established in the age in antiquity. The artist's individuality, originality, sincerity, sensibility, and authenticity in creating a work were considered the only valid standards. As Brown opines, "not since the Renaissance had such a profound change come over the Western consciousness" (Brown, 2001, 10).

While the spirit behind Romantic art was broadly shared, the form and subject matter of Romantic paintings were highly diverse (Vaughan, 2006). Some artists, such as Delacroix, employed classical techniques to depict contemporary subject matter, made all the more dramatic by the use of vibrant colors and thick brush strokes. Most notable is Delacroix's painting *Liberty Leading the People (1830)*, which was intended to celebrate the successful 1830 revolution in France. Likewise, in his *Raft of the Medusa (1918–19)*, Theodore Gericault employed a large-scale historical style to depict a disturbing current event—a shipwreck in which upper class crewman shamelessly abandon women and boatmen to their fates. Other romantic artists, such as Heinrich Fusili and Francisco Goya, explored the depths of their subconscious in their paintings *The Nightmare (1791)* and *The Dream Sleep of Reason Produces Monsters* (1799), while several artists, such as Casper Dieter Friedrick, took landscape painting to a new level, portraying nature as mystical and transcendental (Salmi, 2008, 1985, 53). Seeking to escape the ravages of their time and place, other painters, such as Joshua Reynolds and John Webber, drew inspiration from foreign lands and the Middle Ages.

11.5.2   Expanding Scope beyond French Borders

Contributing to Romanticism's endeavors was its spread throughout Western Europe where its focus varied, depending on the historical context in which it emerged (Rosenthal, 1997, 6; Salmi, 2008, 49; Prickett, 2014). Appearing first in Germany, it was taken up by the Jena intellectuals (Berlin, 1999). Plundered during the Thirty Years War, Germany had succumbed to an inferiority complex especially with regard to France (Berlin, 1999). French pretentiousness had generated a backlash against its lifestyle and neoclassical cultural standards, provoking a German rejoinder building on its Gothic heritage and the German language (Rosenthal, 1997; Salmi, 2008, 29). Whereas French romanticism centered on painting and literature, German Romanticism took on a philosophical bent that, grounded in Pietism, stressed the individual's spiritual life and a personal relation to God (Berlin, 1999, 74–75). Fearing democracy would lead to chaos, as it had in France, German intellectuals looked to Romanticism to uplift (*bildung*) the German population so it might thrive in an integrated, holistic lifestyle. As Beiser (2004) describes, art was believed to have:

> … the power to create an entire world through the imagination. Art could restore moral and religious belief through the creation of a new mythology. It could regenerate unity by 'romanticizing' it, that is by restoring its old mystery, magic, and beauty. (45)

11.5.3   A Conducive Landscape

Given artists' leeway to define their own standards, one might wonder how the Romantic movement ever come about? Recall that movements do not emerge at one stroke. Nor do they appear in one place. Movements follow nonlinear paths according to a trajectory of interactions among like-minded people who share, or develop through ongoing engagement, a congruent vision of a different future. At the turn of the century, Europe was well positioned to cultivate interaction and exchange. Transportation and communication technologies, including the railroads, steam vessels, publishers, newspapers, and periodicals, were increasingly available to meet the cultural demands of a growing, literate population (Blanning, 2012). Artists took advantage of these to ply their works in an ever more, commercial environment. As importantly, publishers and translators assured that major works were accessible throughout Europe (Klancher, 2013). Thus, for example, Jean-Jacques Rousseau's novel *Emile* (1762) and his autobiography *Confessions* (1782) were major best sellers across all of Europe (Berlin, 1999).

11.5.4 Interaction and a Common Vision

Interactions among Romantics were also numerous. For example, Rousseau's writings had a greater influence in Germany than in France. At the same time, the French imbibed romantic ideas from the writings of Madame de Stael, whose review of contemporary German culture, *De L'Allemagne*, was published in 1813. As importantly, the German journal *Antheum*, established by Novalis and the Schlegel brothers in Jena, served as a forum for Romantic thinkers. All the while, artists from many countries met and shared ideas as part of their Grand Tour to Rome (Biome, 2014).

It would seem, therefore, that, much as Granovetter (1995) has argued, Romantics did not need to share common talents or styles of expression to create a movement. What they did needed was to rally around a common, overriding vision. And, despite their differences, there was one thing on which Romantic's did concur. This was the singular role of the artist! Instead of standardizing artifacts, Romantics standardized the image of the artist, spawning thereby a cult, or myth, of the artist as a lone genius, dedicated to his art, and isolated from the materialism and crassness of everyday life. Only the artist could fill the vacuum left by the demise of the institutional arrangements that had governed life up until then (Berlin, 1999; Blanning, 2012; Barker, 2015, 522–523). For Romantics, what was missing from the post-Enlightenment era was a link to the transcendental, the spiritual, and the supernatural. Living in their studios in penury, and isolated from the world around them, they cultivated the idea that only through their introspection and self-reflection could a path to a more utopian life be revealed.

At first glance, the Romantic Movement might be thought to be ephemeral, but its impact has been long term and far reaching (Berlin, 1999). By mid-century, the European landscape had been refashioned. Tastes were changing, and the bourgeoisie was ascendant. In a time of greater prosperity, Romantic symbols and memes no longer resonated with the public. Dependent, now, on the vagaries of the market, artists became more cautious and more responsive to middle-class demands. However, one thing that did not change was the newfound freedom of artists to be true to themselves and to pursue their visions as they saw fit. Art would never be the same.

11.6 Weimar Art

With Germany's defeat in World War One, and the subsequent collapse of the German Empire in November 1919, politics could no longer be ignored. The newly established Weimar Republic was characterized by contention from all sides. As Weitz (2018) described:

> In 1918–23, it was the Left and center; 1924–29, largely the center Right; 1930–33, the authoritarian Right. The first two, at least, demonstrated Weimar's pathologies. Each political configuration, in the end, failed; each fell victim to the concerted attacks of its opponents and its ineptitude. In the end, Weimar's deep seated problems proved far greater than the ability of its leaders to forge consensus and establish hegemony. (83–84)

Unlike the Romantic era, the Weimar Republic teemed with standards, which took the form of signs, symbols, tropes, designs, fashion, and much more. Artists played a critical role in Weimar. Building on the freedom inherited from the Romantics, they employed their rhetorical skills in a wide array of formats. However, in contrast to their predecessors who had eschewed politics, Weimar artists sought to contest and direct the course of a cultural revolution as well as Germany's future.

Although protesting modernity, as had their predecessors, Weimar artists took advantage of all its appurtenances—radio, film, posters, montage, woodcuts, architecture, etc.—to make their case (Bronner, 2012, 14–15, 25). The result was a vibrant burst of creativity, such that all aspects of life came to resemble symbolic, standardized artifacts (Greenberg, 1979, 45; Bronner, 2012, 40). Symbolic art also became a primary mode of contesting German political culture in a battle that became a death match for the fate of Germany (Bronner, 2012).

11.6.1   The Historical Context

From its very beginning, the newly established republic faced untold challenges (Peukert, 1991; Friedrich, 1995; Gay, 1996, 26–27; Weitz, 2018). These encompassed upheavals related to the death of so many German soldiers and the reintegration of the brutally wounded survivors into a bleak, distressed society; the wartime destruction and loss of territories, which crippled recovery efforts; the lack of a governing consensus to support the republic's legitimacy; the emergence of anti-democratic, paramilitary groups; and the financial demands of the Versailles Treaty, which plunged Germany into massive unemployment and uncontrollable inflation.

How were these calamitous events depicted and interpreted for the German public? With the political situation in total disarray, it fell to artists to depict and interpret these events for the German population (Peukert, 1991, 5–6). In this way, German's devastation turned out to be propitious for the flourishing of new art forms (Gay, 1996, 21; Bronner, 2012; Weitz, 2018, 24–25). As Greenberg (1979) describes:

> When during the revolutionary period social and cultural change became a definite possibility, the German artistic-intellectuals revived long cherished goals. Supporting these goals were beliefs in a rational mankind and a belief in the basic psychological similarities in men. During the early months of great excitement, hopes blossomed for the creation of community of individuals in which artist-intellectuals would participate. (79)

11.6.2   The Novembergruppe

Most early Weimar artists—such as Kandinsky, Klee, Marc, and Pechstein—were expressionists, whose style and approach continued that of the pre-war art groups die Brucke (the Bridge) and der Blaue Reiter (the Blue Rider). These artists envisioned themselves as "the bearers of a new humanity", who alone could show the way forward (Bronner, 2012, 36; Weitz, 2018, 25). Their art aimed to provide a symbolic interface to reconcile man to man, and man to society (Greenberg, 1979, 5). Appealing to the subjective aspect of man's being, their efforts were intended to heighten the artistic encounter and generate greater empathy for the downtrodden. Expressionists used line and vibrant color as well as distorted imagery to abstract the "consensual subject" and in the process induce audience participation; intensify awareness; and transcend everyday experience (Bronner, 2012, 32–33, 42). Although the content was diverse, it was typically designed to provoke.

Recognizing the need to act collectively, these artists founded the Novembergruppe in November 1918 (Kohler, 2019). As evidenced in Max Pechstein's illustration, "An alle Kunstler! (To all Artists!)", its aim was to assemble a spectrum of *avant-garde* artists under a broad, inclusive umbrella, allowing them to promote new art (Kohler, 2019, 6). And, in fact, most of Weimar's prominent artists were affiliated with this group (Gay, 1996, 131; Burmeister and Nentwig, 2019, 12).

Notwithstanding its revolutionary stance, the Novembergruppe lacked a common cultural vision and political agenda. While all members were *avant-garde* artists, they leaned in different directions, spanning from the far Left to the Right. To hold itself together, and avoid government censorship, the group remained aloof from party politics. As importantly, its stance reflected the long philosophical tradition in Germany, dating back to Luther and the idealism of Kant, Fichte, and Hegel, which contended that freedom existed only in the sphere of thought (Greenberg, 1979, 3; Laqueur, 2011). As the NovemberGruppe's manifesto stipulated, "the pursuit of radical political aims was a matter for every individual, [and] that the group as an association had merely to pursue radical artistic aims" (Kratz-Kessmeier, 2019, 31). Over time, the Novembergrouppe was reduced to establishing exhibitions where *avant-garde* works were displayed and sold (Kohler, 2019; Kratz-Kessmeir, 2019).

Although the Novembergruppe promoted new forms of art, it failed to achieve its long-term objective—the revolution of culture and society (Burmeister and Nentwig, 2019, 11). The diversity of the group inhibited a common vision and thus collective action. Even more telling, the group lacked political will. As Greenberg (1979) contends:

> Instead of retaining some critical standards *vis a vis* the critical developments of the day, many of the artist intellectuals surrendered these entirely to fantasy building, which they interpreted as their sole role in affecting the real situation. But their utopias were neither acknowledged as such nor made use of to induce striving for a new future. … The artist intellectuals in general left the means whereby to achieve their ends unexplored and unmentioned. Considering their major task, the formation of ideals, of goals towards which men might strive, the revolutionary and reforming artists judged that importance for them lay in "what" and not "how," in symbolic expression rather than instrumentation. (26–27)

11.6.3 Dada

With mounting turmoil in Weimar, some Novembergruppe members denounced its passivity, claiming that, in prioritizing exhibits and sales, the Novembergruppe had been co-opted by the bourgeoisie and become too cozy with the state (Greenburg, 1979, 45; Burmeister, 2019). This consternation was publicly announced in an open letter published in June 1921 in the leftist journal *Der Gegner.* Composed by artist Raoul Hausmann, the letter was signed by George Grosz, Rudolf Schlichter, Hannah Hoch, and Georg Scholz, among others, a number of whom resigned (Burmeister, 2019, 85).

Seeking a confrontational outlet, some dissidents became active in the Dada movement. Believing the "real" revolution had yet to happen, they sought to promote it through their own efforts. Whereas the Novembergruppe focused on the *avant-garde* artistic community, Dada aimed to rid all Germany of its stultifying traditions and revolutionize society as a whole (Greenberg, 1979, 41–46). Dada sought as well to awaken the bourgeoisie from its inertia, so it might resist the systemization and standardization associated with modernity (Greenberg, 1979, 8–10).

Dada was established in Zurich in 1916 during the end of the First World War. Emerging spontaneously through close contacts and interactions among artists and literati at the *Café Voltaire*, Dada provided a venue for protest against the social order that had embraced a horrendous war. Its meaningless label expressed the absurdity of it all (Bronner, 2012; Nentwig, 2019, 37–38). In April 1918, Richard Huelsenbeck transported Dada to Berlin where it gained the support of left-wing journals and artists such as Raoul Hausmann, George Grosz, Wieland Herzfelde, John Heartfield, Franz Jung, and Otto Gross. Believing expressionism deluded the bourgeoisie by distracting their attention with abstractions, Dada members hoped to overturn expressionist standards and, with them, art itself. Hence the slogan "*Die Kunst ist tot*". As described in the 1918 Dada Manifesto:

> For the first time Dadaism has made a break with the aesthetic approach to life by rending all the slogans of ethics, culture, and inwardness, which are cloaks for weak muscles, into their component parts.
>
> *(as quoted in Burmeister and Nentwig, 2019, 39)*

In contrast to expressionists, who worked within the system, Dadaists felt it necessary to destroy the old regime before a new, more humanist one might emerge (Greenberg, 1979, 41). These artists executed their mission through symbolic destruction, as epitomized in their highly satirical, ironic art. As described by Bronner (2012),

> Dada mirrored a world gone mad with its own madness: its pranks, jokes, irony, playful barbarism, rejection of society, celebratory individuals, and expansion of the materials and methods employed by the visual arts all served its protest. (118–119)

Hanna Hoch's photomontage *Cut with the Kitchen Knife through Germany's Last Weimar Beer Belly Cultural Era* stands out in this regard. Although certainly not traditional "art", it served a similar, symbolic purpose (McBride, 2016; Weitz, 2018, 301–303). Using cutouts from local newspapers, journals, and advertisements, Hoch overlaid and juxtaposed them to reflect the period's chaos and mayhem. George Scholtz's painting, *Industrial Farmers* (1920*)*, reflected his experience after the war. It depicts ugly, rapacious farmers seeking to profit from the food shortages and destitute circumstances of the post war trauma. Equally disturbing was George Grosz's painting "*Daum" Marries Her Pedantic Automaton* (Nentwig, 2019, 49).

Although Dada members were highly critical of Expressionists for failing to radicalize the German populace, they were subject to a similar criticism. For Dada remained a closed, elite group, whose satire and irony resonated mainly with its own members. Hence, in 1924, with the stabilization of German currency, and growing prospects for the German economy, Dada began to wane, as its members adopted a more subtle, realist style, *Die Neue Sachlikeit,* that is, The New Objectivity (Gay, 1996, 149–15; Eckmann, 2015; Weitz, 2018, 170–171).

11.6.4 The New Objectivity

Documenting the "real" behind the façade of daily existence, the New Objectivity pulled the curtain on the cultural contradictions of capitalism and the sordid (but to many appealing) underbelly of modernity's frenetic, metropolitan life (Eckmann, 2015, 27–40). As Peter Gay characterizes the period, "culture became less the critic than the mirror of events" (Gay, 1996, 150). A common trope portraying frenzied city life were street scenes of people whirling and mingling through the streets, dancing in night clubs, and frolicking in cabarets. Equally telling were scenes that symbolized Weimar corruption, depicting the very rich and powerful wearing topcoats and hats, smoking cigars, while ignoring or jeering at the poor and destitute. Prostitution was a familiar metaphor used to signal abject poverty, and that anything and everything was up for sale. Paintings of black jazz players and syncopated dancers, as well as futurist paintings of technological artifacts, symbolized Germany's Americanization. At the same time, the blurring of gender roles was rendered in paintings of androgynous women, many times in intimate relationships, with short, cropped hair and dressed in masculine attire. Hostile reactions to the "new German women" were expressed in disturbing paintings of rape/murders.

11.6.5 Bauhaus

Mention must also be made of Berlin's innovative architecture developed by the Bauhaus school (Nentwig, 2019). Even before the war's, architects had promoted a more holistic architectural style uniting painting, sculpture, and architecture. Walter Gropius, leader of the Work Council for the Arts, brought forwarded this idea when, in 1919, he became head of the state Bauhaus in Weimar. Unlike the Dada movement, Bauhaus was constructivist in its orientation, bringing art to the people in a material form (Greenberg, 1979). Gropius wanted to revamp art education by emphasizing craft as the unifying factor and developing a more collaborative relationship between master's and students. He hoped the New Unity of spirit fostered by the Bauhaus could be transplanted to German society (Greenberg, 1979, 56–57). Instead of recoiling from modernity, Bauhaus members embraced it, employing new materials and technologies to enhance their constructions.

Depending on the state for support, the Bauhaus walked a fine line executing its social, humanitarian goals. Many on the right viewed its program as socialist in nature. And the German people, proud of their heritage, were opposed to the use of German taxpayers' money to support a program to modernize Germany's architectural landscape (Greenberg, 1979, 62–64).

Tensions within Bauhaus also fueled its demise. Disagreements focused on organizational standards. While Bauhaus members admired its founder Gropius, and adhered to its overall program goals, some believed that functional artifacts would undermine the spiritual, aesthetic aspects of art. As importantly, although favoring social justice, some Bauhaus members prioritized their own, individual artistic goals over any form of unified approach (Greenberg, 1979, 91).

11.6.6 The Collapse of Weimar and Its Art

Post-war art in Weimar ended with the rise of Nazism. In August 1937, the Nazi Government set up an exhibit of "Degenerative Art". Among the artists displayed were Kandinsky, Klee, George Grosz, Max Beckmann, and Otto Dix, to name a few. All Weimar styles were denounced.

Fearing the worst, many, such as Grosz, Brecht, Beckman, and Weil, immigrated to other countries; others such as Hoch retreated to their internal havens, whereas still others, such as Rudolf Schlichter, collaborated with the Nazis.

Although Weimar artists were inclined to employ their art to improve society, and were prolific in their innovative efforts, they failed to generate a common, consensual standard for that purpose. Even as the November revolution opened up new opportunities and avenues for artistic modes and methods, the subsequent cleavages and conflicts that dominated Weimar politics created fissures in the world of art. As importantly, artistic symbols that intended to create community were incongruent with the times. For, while artists were visionaries, seeking utopia, the German people were far more intent on meeting their daily needs. Thus, as Greenberg has emphasized, "The new age was to remain an ideal, realized only symbolically" (Greenberg, 1979, 25).

11.7 A View from the Whole

What role did standards play in the evolution of European art from the 14th to the 20th centuries? Much as Randall Collins has described (1998), ideas and artistic symbols emerged through interactions and competition among artistic intellectuals. Their diverse interactions engendered bursts of energy and creativity fostering new ideas and artistic styles reinforced by common symbols and standards that defined artistic constellations. At a higher level, these constellations were linked, once again to others, giving rise to new ideas, styles, and standards, providing a font of creative ideas and materials to draw upon and mix and match in new, innovative ways. As importantly, these innovations were codified, and thereby perpetuated, via standards that defined the medium, the subject matter, the level of abstraction, the emphasis on lines, the intensity of color, the thickness of the brush stroke, and the format in which art was presented. These standards were embedded in artifacts, which includes sculptures, paintings, woodcuts, paper montages, film, scripts, dance routines, etc. What about the seeds of reproduction? Enter the artist! It is the artist that bears the standard, rooting it in his or her artworks and art networks, thus making it available to others.

Notes

1 A good example of this is the cultural wars today. As we have seen in the Republican Party for example, cultural issues dominate legislative priorities.
2 The term phase transitions refers to a radical change in structures, as for example when waters turns into ice or steam. See Chapter 5 for an in-depth discussion of phase transitions in historical contexts.
3 For a discussion of humanism as a networked platform, see Chapter 4.

References

Achilles, Manuela (2013) "Democratic Symbols and Rituals in the Weimar Republic," in Channing, Kathleen, et al., eds. *Weimar Publics/Weimar Subjects, Rethinking the Political Culture of Germany in the 1920s*, New York, NY: Berghahn.
Barker, Patrick (2015) *Italian Renaissance in the Mirror*, New York, NY: Cambridge University Press.
Baron, Stephanie (2015) "New Objectivity; German Realism After Expressionism," in Barron, Stephanie, and Sabine Eckmann, eds. *New Objectivity: Modern German Art in the Weimar Republic, 1919–1933*, New York, NY: Prestel.

Barron, Stephanie, and Sabine Eckmann (2015) *New Objectivity: Modern German Art in the Weimar Republic, 1919–1933*, New York, NY: Prestel.

Becker, Howard S. (1982) *Art Worlds*, Berkeley, CA: University of California Press.

Beiser, Frederick (2004) *The Romantic Imperative*, Cambridge, MA: Harvard University Press.

Bell, Daniel (1976) *The Cultural Contradictions of Capitalism*, New York, NY: Basic Books.

Berlin, Isaiah (1999) *The Roots of Romanticism*, Princeton, NJ: Princeton University Press.

Blanning, Tim (2010) *The Culture of Power and the Power of Culture*, New York, NY: Oxford University Press.

Blanning, Tim (2012) *The Romantic Revolution*, New York, NY: Random House, Inc.

Boime, Albert (2014) *Art in the Age of Revolution, 1750–1800*, Chicago, IL: University of Chicago Press.

Boime, Albert (1990) *Art in the Age of Bonapartism*, Chicago, IL: University of Chicago Press.

Bourdieu, Pierre (1993) *The Field of Cultural Production*, New York, NY: Columbia University Press.

Bourdieu, Pierre, and Loic Wacquant (1992) *An Invitation to Reflexive Sociology*, Chicago, IL: University of Chicago Press.

Brown, David Blayney (2001) *Romanticism*, London: Phaidon Press Limited.

Burmeister, Ralf (2019) "Constructivity and Objectivity," in Thomas, Kohler, Ralf Burmeister, and Janina Nentwig, eds. *Freedom: The Art of the Novembergruppe, 1918–1935*, New York, NY: Prestel.

Burmeister, Ralf (2019) "Dada and the Novembergruppe," in Thomas, Kohler, Ralf Burmeister, and Janina Nentwig, eds. *Freedom: The Art of the Novembergruppe, 1918–1935*, New York, NY: Prestel.

Burmeister, Ralf, and Janina Nentwig (2019) "Freedom for Every Pulse, An Introduction," in Kohler, Thomas, Ralf Burmeister, and Janina Nentwig, eds. (2019) *The Art of the Novembergruppe 1918–1935*, New York, NY: Prestel.

Channing, Kathleen, et al. (2013) "Introduction," in Channing, Kathleen, et al. eds. *Weimar Publics/ Weimar Subjects, Rethinking the Political Culture of Germany in the 1920s*, New York, NY: Berghahn.

Collins, Randall (1998) *The Sociology of Philosophies*, Cambridge, MA: Harvard University Press.

Crow, Thomas E (1985) *Painters and Public Life in Eighteen-Century Paris*, New Haven, CT: Yale University Press.

Czikszentmihalyi, Mihalyi (1999) "Implications of a Systems Perspective for the Study of Creativity," in Sternberg, R. J., ed. *Handbook on Creativity*, New York, NY: Harper Collins Publication.

Doherty, Brigid (2018) "The Problem of Politics in Berlin Dada," in Channing, Kathleen, et al. eds. "Introduction," in Channing, et al. *Weimar Publics/Weimar Subjects, Rethinking the Political Culture of Germany in the 1920s*, New York, NY: Berghahn.

Eckmann, Sabine (2015) "A Lack of Empathy: On the Realisms of New Objectivity," in Barron, Stephanie, and Sabine Eckmann, eds. *New Objectivity: Modern German Art in the Weimar Republic, 1919–1933*, New York, NY: Prestel.

Fligstein, Neil, and Doug MacAdam (2012) *A Theory of Fields*, New York, NY: Oxford University Press.

Franceschi, Franco (2008) "The Economy: Work and Wealth," in Najemy, John, ed. *Italy in the Age of the Renaissance, 1300–1550*, New York, NH: Oxford University Press.

Friedrich, Otto (1995) *Before the Deluge: A Portrait of Berlin in the 1920s*, New York, New York: Harper Collins Publishers.

Gardner, Howard (1991) *Creating Minds: An Anatomy of Creativity as Seen Through the Minds of Freud, Einstein, Picasso, Stravinsky, Eliot, Graham, and Gandhi*, New York, NY: Basic Books.

Gay, Peter (1996) *Weimar Culture, The Outsider as Insider*, New York, NY: W.W. Norton and Company.

Goldtwaite, Richard A. (1982) *The Building of Renaissance Florence*, Baltimore. MD: The Johns Hopkins University Press.

Goldtwaite, Richard (2009) *The Economy of Renaissance Florence*, Baltimore, MD: The Johns Hopkins University Press.

Granovetter, Mark (1995) *Getting a Job: A Study of Contacts and Careers*, Chicago, IL: University of Chicago Press.

Greenberg, Allan C (1979) *Artists and Revolution: Dada and the Bauhaus, 1917–1925*, Ann Arbor, MI: UMI Research Press.

Habermas, Jurgen (1991) *The Structural Transformation of the Public Sphere, an Inquiry into a Category of Bourgeois Society*, London, UK: Polity Press.

Harris, Jonathan (2011), "Introduction to Volume II," in Arnold Hauser ed. *The Sociology of Art*, vol. 2, Abingdon, Oxon, Routledge.

Hauser, Arnold (2011) *The Sociology of Art*, vol. 2, Abingdon, Oxon: Routledge.

Hibbert, Christopher (1974) *The House of Medici: Its Rise and Fall*, William Morrow Paperbacks.

Hohler, Thomas (2019) "The Art of the Novembergruppe from 1918–1935," in Kohler, Thomas, Ralf Burmeister, and Janina Nentwig, eds. *Freedom: The Art of the Novembergruppe, 1918–1935*, New York, NY: Prestel.

Holland, John (2014) *Signals and Boundaries*, Cambridge, MA: MIT Press.

Honour, Hugh (1979) *Romanticism*, New York, NY: Westview Press.

Hutter, Michael (2015) *The Rise of the Joyful Economy, Artistic Invention and Economic Growth from Brunelleschi to Murakami*, New York, NY: Routledge.

Jenkins, Richard (2002) *Pierre Bourdieu*, New York, NY: Routledge.

Joost-Gaugier, Cristine (2013) *Renaissance Art, Understanding Its Meaning*, Maldon, MA: John Wiley & Sons.

Kent, Dale (2008) "The Power of the Elites: Family, Patronage and the State," in *John Najemy, ed. (b) Italy in the Age of the Renaissance, 1300–1550*. New York, NY: Oxford University Press.

Klancher, Jon (2013) *Transfiguring the Arts and Sciences, Knowledge and Cultural Institutions in the Romantic Age*, Cambridge: Cambridge University Press.

Kohler, Thomas, Ralf Burmeister, and Janina Nentwig, eds. (2019) *Freedom: The Art of the Novembergruppe, 1918–1935*, New York, NY: Prestel.

Kratz-Kessmeier, Kristina (2019) "Paving the Way for Modernism—The Novembergruppe's Role in Shaping Art Policy During the Weimar Republic," in Kohler, Thomas, Ralf Burmeister, and Janina Nentwig, eds. *Freedom: The Art of the Novembergruppe, 1918–1935*, New York, NY: Prestel.

Kristeller (1990) *Renaissance Thought and the Arts*, Princeton, NJ: Princeton University Press.

Laqueur, Walter (2011) *Weimar: A Cultural History*, New Brunswick, NJ: Transaction Publishers.

Lowy, Michel, and Robert Sayre (2001) *Romanticism Against the Tide of Modernity*, Durham, NC: Duke University Press.

McBride, Patrizia C. (2016) *The Chatter of the Visible: Montage a Narrative in Weimar Germany*, Ann Arbor, MI: University of Michigan Press.

Maxson, Brian Jeffrey (2014) *The Humanist World of Renaissance Florence*, New York, NY: Cambridge University Press.

Muir, Edward (2008) "Representation of Power," in Najemy, John, ed. *Italy in the Age of the Renaissance, 1300–1550*, New York, NY: Oxford University Press.

Najemy, John M. (2008a) *History of Florence 1200–1575*, Maulden, MA: John Wiley and Sons, Ltd.

Najemy, John, ed. (2008b) *Italy in the Age of the Renaissance, 1300–-1550*, New York, NH: Oxford University Press.

Nentwig, Janina (2019) "A Strong Will for New Architecture," in Kohler, Thomas, Ralf Burmeister, and Janina Nentwig, eds. *Freedom: The Art of the Novembergruppe, 1918–1935*, New York, NY: PRESTEL.

Nentwig, Janina (2019) "Liberating Energies of the New Art," in Kohler, Thomas, Ralf Burmeister, and Janina Nentwig, eds. *Freedom: The Art of the Novembergruppe, 1918–1935*, New York, NY: Prestel.

Nentwig, Janina, and Peters, Olaf, ed. *Berlin Metropolis, 1918–1933*, New York, NY: Prestel.

Padgett, John F., and Walter W. Powell (2012) *The Emergence of Organizations and Markets*, Princeton, NJ: Princeton University Press.

Peter Gay (1995) *The Enlightenment, The Rise of Modern Paganism*, New York, NY: W.W. Norton.

Peters, Olaf (2015) "Art, Culture and Politics Between the Wars," in Peters, Olaf, ed. *Berlin Metropolis, 1918–1933*, New York, NY: Prestel.

Peters, Olaf, ed. (2015) *Berlin Metropolis, 1918–1933*, New York, NY: Prestel.

Peukert, Detlev J. K (1991) *The Weimar Republic: The Crisis of Classical Modernity*, London: Penguin Press.

Prickett, Stephen, ed. (2014) *European Romanticism, A Reader*, New York, NY: Bloomsbury.

Quillen, Carol Everhart (2008) "Humanism and the Lure of Antiquity," in Najemy, John, ed. *Italy in the Age of the Renaissance, 1300–1550*, New York, NH: Oxford University Press.

Rosenthal, Leon (1997) *Romanticism*, New York, NY: Parkston International.

Safranski, Rudiger (2014) *Romanticism, A German Affair*, Evanston, IL: Northwestern University Press.

Salmi, Hannu (2008) *19th Century Europe, A Cultural History*, Cambridge: Polity Press.

Sawyer, Keith (2006) *Explaining Creativity: The Science of Human Innovation*, Oxford: Oxford University Press.

Scott, Hamish, and Brendan Simms, eds. (2007) *Culture of Power in Europe During the Long Eighteen Century*, Cambridge: University of Cambridge Press.

Vaughan, William (2006) *Romanticism and Art*, New York, NY: Thames and Hudson Ltd.

W. Mitchell Waldrop, (2005) *Complexity, The Emerging Science at the Edge of Order and Chaos*, New York, NY: Open Road.

Walsh, Linda (2017) *A Guide to Eighteenth Century Art,* West, Sussex: John Wiley & Sons, Inc.

Weiner, Eric (2016) *The Geography of Genius, Lessons from the World's Most Creative Places*, New York, NY: Simon and Schuster.

Weitz, Eric D (2018) *Weimar Germany, Promise and Tragedy*, Princeton, NJ: Princeton University Press.

Woods, Kim, ed. (2012) *Art and Visual Culture 1100–1600: Medieval to Renaissance*, London: Tate Publishing in Association of Open University.

Zorzi, Andrea (2008) "The Popolo," in Najemy, John, ed. *Italy in the Age of the Renaissance, 1300–1550*, New York, NH: Oxford University Press.

PART IV

Standards and Complexity

Summing It Up

12

STANDARDS, EMERGENCE, AND COMPLEXITY—PIECING THE PUZZLE TOGETHER

12.1 Introduction

Formulations of complexity have recently taken hold as an innovative way of thinking about phenomena from realms embracing the biological, technical, economic, social, cultural, and the geographic (Sawyer, 2005; Haynes, 2018; Daems, 2021; Beggs, 2022). Such attention is not surprising. By linking analyses across the vast landscape of our knowledge, complexity analyses cannot only greatly expand the research agenda; as importantly—much as complexity theory portends—these analyses will also benefit from the integration and agglomeration of diverse resources and ideas (Bunge, 2003).

Embracing this conceptualization of complexity, this concluding chapter demonstrates how a standards-based analysis of complexity sheds light on many issues that have hitherto suffered from truncated analyses due, in large part, to their failure to fit squarely into traditional disciplinary silos. This concluding chapter seeks to fill this gap by characterizing standards as the linchpin of complexity, and hence the "missing link" between cause and effect. It argues that, when standards are linked together over time and space in a networked fashion, they provide a road map for identifying the critical points in a sequence of events at which standards might be reconfigured so as to alter outcomes.

Notwithstanding, and perhaps because of, complexity's relevance across disciplines, it has been difficult to define, and conceptualize, this concept. Given diverse analytic approaches, different aspects of complexity are pronounced in some accounts, while minimized in others. Thus, instead of theorizing complexity, most analysts have, more often than not, simply characterized it in terms of its generally accepted attributes. Clearly, more theoretical work needs to be done! The failure to conceptually link complexity's attributes together, so as to account for their convergent impacts, weakens scholars' ability to employ complexity analysis on behalf of problem solving. As John Holland (1995) opines:

> Theory is crucial. Serendipity may occasionally yield insight, but is unlikely to be a frequent visitor. Without theory, we cannot separate fundamental characteristics from fascinating idiosyncrasies and incidental features. Theory supplies landmarks and guideposts, and we begin to know what to observe and where to act. (5)

DOI: 10.4324/9781032721125-17

Pursuing the goal of theory building, this chapter first characterizes standards and defines their role in delineating and channeling interactions. Second—as laid out in Sections 3 through 8, it describes not only the key attributes associated with complexity, and specifies how standards determine them, integrates them within a coherent, dynamic whole. Included among these attributes are size and heterogeneity, non-linearity and holistic outcomes; self-organization for greater fitness; hierarchies and platforms; linchpins of emergence; adaptation and small world platforms; and fractal societies with standardized constituents. Third, sections nine through twelve draw some overarching conclusions, characterizing how a *theory of complexity, driving by standards,* cannot only inform and enhance our efforts at policy analyses but also transcend our existing, disciplinary divides.

The following discussion is derived from the theoretical literature relating to complexity as well as the empirical analyses put forth in the preceding chapters.

12.2   Standards—The Building Blocks of Complexity

As argued in Chapter 1, "The Introduction", standards—given their vast range—can best be defined, generically, as interfaces that link entities be they between individuals, machines, or elements of the natural world. Such entities are multitudinous: they might include, for example, individuals, groups, words, currencies, technologies, ideas, cultural artifacts, academic disciplines, professions, nation states, etc. In connecting these diverse entities, so they can communicate and engage with one another, standards assume a wide variety of forms. These include, for instance, signs, signals, symbols, codes, grammar, norms, conventions, tropes, cues, pheromones, and stereotypes, to name but a few. Standards not only tie entities together; they also provide the rules governing their interactions. That is to say, they determine how, why, and by virtue of whom/what interactions take place, as well as the mode of interactions, the conditions under which interactions occur, and the appropriateness of interactions. As the growing literature on networks makes clear, these rules determine the configuration of complex entities, along with their significant consequences (Buchanan, 2002; Burt, 2007; Borgatti, et al., 2008; Easley and Kleinberg, 2010; Skyrms, 2014). Hence, by establishing the boundaries linking complex entities, and determining how they relate to one another, standards function as the drivers of complexity, serving as the building blocks of complex adaptive systems (Holland, 2014).

To grasp how standards function in these regards, we might consider how, as described in Chapter 6, "Standards and Evolution in a Complex World", pheromones both link, and direct, the life of resident ants living within a complex, adaptive community. Ant communities are self-organized (Johnson, 2012, 29–34; Theise, 2023, 26–27). That is to say, instead of following a leader, ants track pheromone signals that they, and others, leave in their paths. Thus, an ant might deposit a pheromone trail when seeking food, both to locate the source for other ants as well as to record its own route back to its nest. As other ants read these signals, they too follow the path to food, reinforcing it along the way. Intersecting with one another, ants adapt to additional pheromone signals that provide updates as to whether there is a need for additional food, or whether they should return to their nest to assume other roles (Johnson, 2012, 29–34; Theise, 2023, 27–28).

12.3   Size and Heterogeneity

Complexity emerges exponentially when populations, of any kind, increase both in their size and heterogeneity, engendering greater and more elaborate interactions among them. Thus, viewing history from the *longue durée,* we can trace the rise of complexity in society concomitantly with the growth in population. As Malthus argued in his 1789 work, *Essays on the Principle of Population,* population growth is exponential, other things being equal. Moreover, heterogeneity accompanies population growth due to the specialization of functions and the division of labor (Durkheim, 1997; Smith, 2013).

Standards play a critical role in determining both variables. As we have seen, the size and scope of any network is a function of its internal and external interactions. Standards provide the means for such interactions to take place. Standards also determine the division of labor. As detailed in Chapter 10, "Crafting Identity with Standard Memes and Symbols", by taking the form of roles, standards not only identify, and bind, diverse entities; they also distinguish among them by delineating the functions and expectations associate with each. This phenomenon holds true for technologies as well, as detailed in Chapter 8, "Standards, Modularity, and Innovation". As described therein, when technological modules are fitted with standard interfaces, they can be combined and recombined in innovative ways so as to produce an ever-expanding array of innovative technologies. Not surprisingly, biological systems exhibit similar characteristics. As ant biologist Nigel Franks has described, a small gathering of ants will wander around unwittingly until they die, whereas the assembly of a massive group, directed by standardized pheromones, will operate as a "superorganism" with "collective intelligence" (Mitchell, 2009, 21). In all these cases, the standards governing these entities breed ever-increasing complexity by generating greater, and more elaborate, interactions (Holland, 2014, 5).

12.4   Non-Linearity and Holistic Outcomes

Building upon the mechanical philosophy of the 17th century, most researchers—especially those in the physical sciences—have adhered to a "reductionist" mode of analysis, which posits that, to fundamentally understand a phenomenon, one must begin by analyzing its individual parts, and then, on that basis, deduce the nature of higher level structures (Bunge, 2003, 23; Kauffman, 2005). Their approach assumes that events, when carried out in a process, are linear, so that outcomes can be anticipated and explained by simply "adding up" the steps in their development. As complexity theory contends, such analyses fail to acknowledge the intervening variables—that is to say, the collective structures between cause and effect. As a result, reductionists cannot account for the likelihood that, when events are integrated in non-linear ways, their outcomes will be greater than the sum of their parts (Kauffman, 1995; Mayfield, 2013). Nor can they identify the critical points in a system, which—if reconfigured—might alter the course of emergent outcomes. As we have seen throughout, standards play an essential role in both cases. Two highly illustrative examples are laid out below.

12.4.1   Consider Packet Switched Networks

To grasp the role of standards in facilitating non-linearity, consider packet switching, the technological foundation of the Internet. Developed in the sixties and the seventies, packet switching was first deployed in 1981, when the Defense Department's released ARPANET, a network connecting its associated universities and research centers. In contrast to circuit switching—the prevailing system at the time— packet switching is a bottom-up communication process that employs the internet's TCP/IP standards to dynamically distributes messages in real time over a digital network, depending on whichever channels are available at the moment. To do so, messages are broken up into small, digital bits and then enfolded into several packets that travel in parallel across the network. Each packet contains "a header"—that is, digital bits of information—which directs the packet to its final destination where the bits from all corresponding packets are reconfigured to create the original message. Along the way, the packets are stored and forwarded at switching nodes, such as routers, where they are once again assigned in a sequential fashion to the best available routes to their final destinations. By dynamically channeling communications throughout the network, standards perform an essential function in this non-linear process (Abbate, 1999; Clark, 2018).

12.4.2   Creating the "Whole" through Non-linear Social Processes

The role of standards in generating non-linear processes is not limited to technical systems; standards serve the same function in the social systems as well. In both instances, communications based on standard rules determine how, and to what effect, non-linear activities take place. Recall the workings of jazz ensembles and improvisation theater as depicted in Chapter 3, "Standards, Norms, and Emergence in complex Settings". In both cases, performances start out from scratch, with no prior scheme or direction. It is standards that pave the way forward in a non-linear process. Facilitated and constrained by a minimal set of standards and constraints, the theme, or plot, emerges collectively through the interactions among the multiple participants. The results of these performances are not only full of surprises; they are also greater than the sum of their parts (Sawyer, 2005, 4). Referencing the importance of standards, Sawyer (2008) opines:

> Participants must be able to innovate, but to do so they must build upon shared conventions and repertoires, which—like a shared language—allow group members to lead and follow one another as in a unified whole. (5)

12.5   Self-Organizing for Greater Fitness

Complexity science posits that self-organization takes place when simple entities, operating according to basic rules generated over time within their internal and external environments, organize themselves in hard-to-predict complex ways, so as to enhance their fitness (Kauffman, 1993; Holland, 1999, 7). Employing very simple rules embedded in computer codes, computer scientists have generated such self-organizing systems, as in the case of cellular automata and genetic algorithms (Holland, 1999; Kauffman, 1995; Waldrop, 2005). For example, John Holland's genetic algorithm performs a fitness selection function by comparing the fitness value of diverse populations depending on how they perform a given task.

Standards-based self-organizing systems are not limited to computer programs; they are prevalent in the social and ecological arenas as well. For example, Chapter 2, "Assent up the Fitness Landscape: How the West Was Won", documents how the American Pioneers, while making their way West, developed *en route*, and in the course of their interactions, new standards adapted to their ever-changing environments. With each new set of standards, the Pioneers enhanced their fitness by gaining the wherewithal to operate successfully in one locale before advancing to the next. We might say, then, that, based on these evolving standards, the western migration followed the path of the "adjacent possible", a term developed by complexity biologist Stuart Kauffman to refer to adjoining possibilities in the fitness landscape.

As described in Chapter 10, "Crafting Identity with Standard Memes and Symbols", standards also enhance fitness by engendering and reinforcing communities of interaction that function as platforms for emergence. Building on the work of Randall Collins, this chapter contends that communities of individuals emerge through a process of interaction based on symbolic, standard-based, ritual chains. In group contexts, mutual focus of attention occurs, and emotional entrainment builds up, so "that self-generating feedback processes give rise to moments of gripping emotional experience". These experiences act like magnets of cultural significance where "culture is created, denigrated, or reinforced" (Collins, 2005, 12–13). Much as is the case of the merged technological components described in Chapter 8, such rituals are transformative in that they "take some emotions as ingredients and transfer them into other emotions", so as to fuel concerted group action (Collins, 2005, 13).

The narratives documented in Chapters 6 through Eleven provide corroborating evidence of this process. Chapter 6, "Standards and Evolution in a Complex World", describes for one, how plants employ chemical signals to generate "community" responses to potential dangers. Chapter 7, "Complex Monetary Outcomes: Winners and Losers, and the Standards Determining Them", describes how money—defined as a standard of value—not only engenders platforms of economic activity, which increase over time in size and complexity, but also—in the process—reconfigures relationships, giving rise to new winners and losers. Chapter 8, "Standards, Modularity, and Innovation", characterizes how standards, taking the form of technical interfaces, link modular technologies together not only to create new, innovate technologies, but also to provide the resources and information needed to exploit them. In turn, Chapter 9, "How Standards Engender Trust", details how societal standards, serving as norms, conventions, practices, symbols, protocols, etc., converge to form platforms of trust, upon which standards are integrated and reinforced so as to generate emergent, social capital. Likewise, Chapter 10, "Crafting Identity with Standard Memes and Symbols", recounts how identities are formed in an iterative, mutually reinforcing fashion, when standardized role behavior and role expectations converge. Lastly, Chapter 11, "The Artist as Standard Bearer", describes how European artists, from the period of the Enlightenment to that of the Weimar Republic, came together in coherent ensembles based on common artistic standards to employ standards as cultural memes intended to rally politicians and the public around their social and political points of view.

12.6   Platforms and Hierarchies Up the Fitness Landscape

The concept of complex hierarchies dates to the time of Herbert Simon's 1962 paper, "The Architecture of Complexity", which characterizes complex systems as being composed of subsystems. More recently, complexity scientists have portrayed these subsystems as hierarchical building blocks that lead upwards to greater fitness and complexity (Kontopoulos, 1993, 180–207; Kauffman, 1995; Mitchell, 2023, 127–128). According to Bunge, for example, these building blocks emerge from multiple interactions that lead to the integration and radical restructuring of their components (Holland 1999, 11; Bunge, 2003, 27–30; Elder-Vass, 2010, 30–32). As Holland contends, these subsystems are bound together, and governed, by "tags", which I characterize as standards in their many forms[1] (Holland, 1999, 11–13).

In Chapter 4, "Platforms: Springboards for Evolutionary Outcomes", I conceive of these building blocks as *platforms,* where networks of diverse interactions converge in a multiplexed fashion so that actors/actants, as well as their purposes and functions, align. Common to many platforms is their *small world architectures,* as defined by the standards that govern and link them to their external environments. These standards provide an infrastructure that furnishes both a place, and a time period, for new resources and diverse information to be agglomerated and exchanged. Drawing on these assets, platform actor/agents generate and reorganize themselves around new standards and practices that give rise to emergent outcomes, allowing them to enhance their fitness and adapt to their changing circumstances (Elder-Vass, 2010, 33–34). In this sense, platforms are similar to biological *niches* in that they constitute layers in a hierarchical, or heterarchical, configuration, providing stepping stones up and down the evolutionary fitness landscape.

12.7   The Linchpin of Emergence

The complex processes by which platforms are generated and layered atop one another in a hierarchical, or heterarchical fashion, engenders *emergence*. As we have seen in Chapter 3, "Standards, Norms, and Emergence in Social Settings", emergence is fueled by changing standards configurations. Emergence occurs when two or more entities are integrated together, such that their amalgamation is not only decidedly different from, but also more complex than, the sum of their parts. As importantly, emergence is the means by which actors move up the fitness landscape (Kontopoulos, 1993; Kauffman, 1995; Holland, 1999; Bunge, 2003; Johnson, 2012).

Although the term "emergence" is much in vogue today, it is not a new idea. Intellectuals dating back to Herbert Spencer, Adam Smith, Charles Darwin, Friedrich Engels, and Alan Turing have pondered this phenomenon, without actually accounting for how it comes about (Johnson, 2012, 18–19). Most elusive has been the task of identifying and characterizing the precise *mechanisms* by which emergence takes place (Bunge, 2003, 39–40; Bell, 2024). Although computer scientists have employed digital models such as cellular automata and genetic algorithms to demonstrate emergence,[2] they have yet to develop a theoretical explanation for how rules themselves emerge, as well as the means by which the *mechanisms,* inherent in these rules, determine diverse effects. (Mayfield, 2013). Also lacking from these efforts, as well as in many biological evolutionary explanations—as detailed in Chapter 6—is a characterization and explanation of the role that contingent factors, and intervening variables, play in determining results (Mayfield, 2013, 49, 50–51).

The chapters throughout this book illustrate how standards, in whatever their forms, serve as interfaces that embody rules of interaction between individuals, machines, or elements of the natural world. As such, standards determine who and what can interact, the modes of interaction, the conditions under which interactions take place and the appropriateness of interactions. Based on this definition, standards embody a purpose insofar as they provide instructions, as well as a criterion, by which to measure outcomes. As Mayfield attests, these instructions aim to achieve specific outcomes, which always entail "being for something or to fit some local circumstance" (Mayfield, 2013, 51).

Emphasizing how standard "instructions" lead to greater fitness through emergent processes, Chapter 3 documents a wide array of empirical examples. Included are descriptions of the emergence of complexity theory at the Santa Fe Institute; the emergence of a *modus operandi* between Jews and Arabs at the Arava Institute in the Negev desert; the emergence and unfolding of music and drama in jazz and improvisation productions; the emergence of the Renaissance in 13th-century Italy; and the emergence of global cities today. These diverse cases all illustrate how, due to the adaptation of standards through interactive processes, actors/agents are able to leap from one layer of the complexity hierarchy to the next, creating thereby a path for social evolution.

12.8   Adaptations and Small World Platforms

Adaption takes place when agents, or actors, reconfigure their organizational structure so as to prosper and prevail in a new environment defined by the standards constituting the evolving fitness landscape (Holland, 1999). As in the case of the jazz ensembles described in Chapter 3, and the historical narratives presented in Chapter 5, "Phase Transitions in the Middle Ages", adaptation is an on-going process in which agent/actors adapt to new circumstances by generating new standards of interaction, creating thereby a pathway to a greater "whole". As set forth in Chapter 4, "Platforms: Springboard for Evolutionary Outcomes", platforms provide an infrastructure that furnishes space, as well as a time period, for the agglomeration of new resources and diverse information. Drawing on these assets, platform actors, or agents—as the case may be—generate new standards and practices that allow them to adapt to changing circumstances.

As Chapter 4 describes, the attribute that makes platforms adaptive is their *small world* configuration as determined by the standards that constitute their architectures (Buchanan, 2002). Bounded and regulated by standards, small worlds entail a central core, or cluster, made up of strong ties, together with weak ties, which extend outward from the core to the environment. As illustrated in Chapter 9, "How Standards Engender Trust", strong ties are essential for facilitating the trust and collaboration required for information exchange and collective action (Burt, 2007). But, as we have seen in Chapter 4, strong ties tend to inhibit adaptation and reinforce the status quo. But weak ties, which extend beyond a platform, are ideal for retrieving diverse resources and novel, up-to-date information. When strong and weak ties are joined in a unified, small world platform, interactions are not only highly adaptive but also, at one and the same time, the source of emergent outcomes.

12.9   Fractal Societies with Standardized Constituents

In 1982, polymath Bernoit Mandelbrot coined the term "fractal", to describe objects that, unlike uniform Euclidean geometric shapes, such as spheres, cones, circles, and straight lines, take on irregular forms. As he argued in his book, *The Fractal Geometry of Nature,* fractal objects are far more common than previously assumed. Although Swedish mathematician Helge von Koch identified the fractal nature of snowflakes as early as 1904, scientists paid little attention due to their seeming irrelevance as well as the limitations of mathematics at the time (Linton, 2019). As Gerhard Schroeder (2009) notes:

> Yes, for all these years, we *have* been living with fractal arteries, not far from fractal river systems draining fractal mountain-scapes under fractal clouds, toward fractal coastlines. But, kin to Moliere's would-be gentleman, we lacked the proper prose—fractal noun and adjective—that Benoit B. begot. (xvi)

A common characteristic of all fractals is their self-similarity at all scales (Schroeder, 2009; West, 2018). Much as Russian dolls are ordered, the largest fractal contains within it increasingly smaller fractals, each housing smaller fractals, which—although individual and unique—are replicated at every scale with respect to their organizational structures and functions. Thus, we might say that, although the whole is greater than the sum of its parts, from a structural perspective, we can see that each part mirrors the whole.

Because the analysis of fractals is based on mathematics, it is not surprising that its applicability has been limited for the most part to the hard sciences, where a primary goal has been to develop mathematical proofs. However, missing from these analyses is an attempt to apply the concepts underlying the phenomenon of fractals to social systems. Much like the early mathematicians who lacked the analytical tools to delve into the intricacies of fractals, so today, many social scientists, seeking to avoid organic, system-wide thinking and functional analysis, have missed an opportunity to employ the lens of fractals to account for the consistent, self-replicating patterns of structural interactions and non-linear processes giving rise to evolutionary emergence.

12.10   Standards, the Building Blocks of Evolution

Differences between sociological and biological accounts of evolution have a long history, dating back to mid-19th century, when August Comte and Emile Durkheim —the founders of sociology—established their positivist intellectual agenda, setting it firmly apart from the ideas of both the theological philosophers and material scientists of the day. As laid out in Chapter 6, notwithstanding Herbert Spencer's efforts to unite these perspectives within his theory of Social Darwinism, these cleavages have persisted, only to be exacerbated with the rise of the field of Sociobiology, which seeks to account for social behavior based on the evolution of individual genes.

As described in Chapter 6, a major barrier to resolving these differences stems from the fact that each of these analytical approaches not only focuses on a different unit of analysis but also fails to effectively link them together. While population geneticists attribute evolution to the random, individual genotype, social scientists characterize evolution in terms of the collective, purposeful behavior of the phenotype. The chapter argues that one way of reconciling the diverse paradigms is to view evolution through the lens of complexity theory and the role that standards play in both generating and governing evolutionary outcomes. Because genotypes and phenotypes both interact based on standards, whether in the form of codes, signals, norms, or conventions, one way of achieving a unified, evolutionary perspective is to focus on *standards* in whatever their form—as the unit of analysis.

12.11   Leveraging Standards to Solve Problems and Channel Outcomes

A key concern among social scientists relates to the question of individual agency. Or, more precisely, to what extent can individuals control the effects of their actions, notwithstanding a myriad of structural constraints? At one extreme are those social scientists who, employing a theoretic framework based on methodological individualism, believe individuals have considerable leeway to effectuate their choices. At the other extreme are those who argue that individual agency is severely limited, given society's powerful structural constraints[3]. Characterizing the difference between these two perspectives, Bunge (2003) notes:

> Whereas individualism stresses action and underrates and even ignores social ties, holism underrates agency and overrates bonds. In other words, while individualists take individual behavior for granted and regard it as the source of social patterns; holists take the later for granted and regard it as the source of individual behavior. Neither regards the outcome—the behavior or structure, as the case may be—as problematic. (130)

Although scholars from the 1970s—such as Bhaskar, Giddens, and Bourdieu—have sought to overcome this dichotomy by conceiving of agency and structure as being inherently intertwined, they have failed to address the micro-macro link—that is, how, and by what means, actors exercise agency as they negotiate their way through the structural hierarchy of the fitness landsape (Kontopoulos, 1993, 211; Bunge, 2003, 91).

As the chapters in this volume illustrate, complexity theory identifies a pathway for achieving agency whether at the individual, group, or national level. Drawing on this theory, and as spelled out in Chapters 4 and 7, I contend that the social world, as well as many other worlds, are comprised of platforms, which are not only layered upon one another in a hierarchical fashion but also—as in the case of fractals—mirror one another at different scales. Most importantly, I make the case that it is standards, in whatever their form, that function as the linchpins of complexity, both channeling interactions within and across system platforms, while agglomerating the system's components, and binding them together.

By identifying and focusing on critical standards, and how and by whom they are determined, it is possible to gain greater agency and, hopefully, better leverage with respect to our futures. Just as standards can be altered to determine the trajectory of the packet switched networks described above, so too we can employ standards that govern interactions, to redirect events in new—and hopefully—more promising directions.

This conclusion, which is drawn from the present analysis, mirrors one I learned over the years while conducting studies at the Office of Technology Assessment. In contrast to technology determinists, who draw a straight line between technologies and their outcomes (much as neo-Darwinians do today with respect to the evolution of genes), I found that the intervening variables comprising the context in which technologies are deployed and diffused are equally, if not more, significant than the technologies themselves. It is, in fact, based on this understanding that I have characterized these intervening variables as standards, which I claim to be the linchpin and leverage points for evolutionary change.

12.12 Implications for the Social Sciences

Years ago, when presenting an overview of my well-received OTA report *Critical Connections: Communications for the Future,* to a group of academics in Canada, I was stunned when someone from the audience called me out for being a "structural functionalist". It should not have surprised me. Having received my advanced education at Columbia University, which maintained an institutionalist approach in the face of trending quantitative analysis, I had been rigorously schooled in the works of Durkheim, Parsons, Weber, and Merton. And, despite the subsequent shift towards post-modernism and away from grand narratives, I continue to view the world in structural terms. In fact, it was this focus that attracted me to complexity theory.

Most notably, complexity, as I have characterized it, is not bound by analytic disciplines; it is inherent in all realms of life. Moreover, in each realm, complexity is derived from interactions that are governed by standards. Insofar as standards are inherent is all complex activities, we can, by employing a complexity framework based on standards, transcend the many divisions within the social sciences, and thereby enrich our research agendas as well as adapt our focus to deal with unforeseen, unpredictable outcomes. As Bunge has maintained, because problems are macrosocial and multidimensional, solving them cannot be reductionist; it must be integrationist as well (Bunge, 2003, 187). With a focus on standards, complexity theory, offers a unique way doing this. Above all, it (1) generates innovative insights about social phenomenon; (2) identifies leverage points that might be employed to affect social outcomes; and (3) unifies the social sciences in a holistic, interdisciplinary fashion.

Notes

1 As Holland notes, tags—much like headers in a packet switched network—are the means by which interactions are aggregated; channeled through the network; and bounded together. However, they are not fixed in time (Holland, 1999).
2 To do so, they have embedded "simple rules" in computer programs in the form of if/then statements, and then traced the results. By positing diverse rules in a top-down manner, these scientists have been able to make real time comparisons about how outcomes change in accordance with changing rules (Marro, 2014, 4–9).
3 Characterizing the difference between individualism and holism, Bunge notes, "Whereas individualism stresses action and underrates and even ignores social ties, holism underrates agency and overrates bonds. In other words, which individualists take individual behavior for granted and regard it as the source of social patterns; holists take the later for granted and regard it as the source of individual behavior. Neither regards the outcome—the behavior or structure, as the case may be—as problematic. Let along as one sides of one and the same social coin" (Bunge, 2003, 130).

References

Abbate, Janet (1999) *Inventing the Internet*, Cambridge, MA: MIT Press.
Arthur, W. Brian (2009) *The Evolution of Technology: What It Is and How It Works*, New York, NY: Free Press.
Beggs, John M. (2022) *The Cortex and the Critical Point: Understanding the Power of Emergence*, Cambridge, MA: MIT Press.
Bell, Philip (July 10, 2024) The New Math of How Large-Scale Order Emerges, *Quanta Magazine.*
Borgatti, Stephen P., Martin G. Everett, and Jeffrey C. Johnson (2008) *Analyzing Social Networks*, New York, NY: SAGE Publications.
Bryne, David, and Callaghan (2023) *Complexity Theory and the Social Sciences: The State of the Art*, New York, NY: Routledge.
Buchanan, Mark (2002) *Small Worlds: The Groundbreaking Theory of Networks*, New York, NY: W.W. Norton and Company.
Bunge, Mario (2003) *Emergence and Convergence: Qualitative Novelty and the Unity of Knowledge*, Toronto, Canada: University of Toronto Press.
Burt, Ron S. (1992) *Structural Holes: The Social Structure of Competition*, Cambridge, MA: Harvard University Press.
Burt, Ron (2007) *Brokerage and Closure: An Introduction to Social Capital*, New York, NY: OUP Oxford.
Clark, David (2018) *Designing the Internet*, Cambridge, MA: MIT Press.
Collins, Randall (2005) *Interaction Rituals Chains*, Princeton, NJ: Princeton University Press.
Daems, Dries (2021) *Social Complexity and Complex Systems in Archaeology*, New York, NY: Routledge.
Durkheim, Emile (1997) *The Division of Labor in Society*, Trans. W.D. Halls, New York, NY: Free Press.
Easley, David, and Jon Kleinberg (2010) *Networks, Crowds, and Markets: Reasoning about a Highly Connected World*, New York, NY: Cambridge University Press.
Elder-Vass, Dave (2010) *The Causal Powers of Social Structures: Emergence, Structure, and Agency*, New York, NY: Cambridge University Press.
Haynes, Philip (2018) *Social Synthesis: Finding Dynamic Patterns in Complex Social Systems*, New York, NY: Routledge.
Holland, John (1999) *Emergence, From Chaos to Order*, New York, NY: Basic Books.
Holland, John (2014) *Signals and Boundaries: Building Blocks for Complex Adaptive Systems*, Cambridge, MA: MIT Press.
Johnson, Steven (2012) *Emergence: The Connected Lives of Ants, Brains, Cities, and Software*, New York, NY: Scribner Book Company.
Goffman, Erving (1959) *The Presentation of Self in Everyday Life*, New York, NY: University of Edinburgh Social Sciences Research Center, Anchor Books Edition.

Kauffmann, Stuart (1995) *At Home in the Universe: The Search for Laws of Self- Organization and Complexity*, New York, NY: Oxford University Press.

Kontopoulos, K. M. (1993) *The Logics of Social Structure*, Cambridge; New York, NY: Cambridge University Press.

Larson, Edward J. (2006) *Evolution: The Remarkable History of a Scientific Theory*, New York, NY: Random House.

Linton, Oliver (2019) *Fractals on the Edge of Chaos*, Glastonbury, Somerset: Wooden Books Ltd.

Marro, Joaquin (2014) *Physics, Nature and Society: A Guide to Order and Complexity in Our World*, New York, NY: Springer.

Mayfield, John E. (2013) *The Engine of Complexity, Evolution as Computation*, New York, NY: Columbia University Press.

Mitchell, Melanie (2009) *Complexity: A Guided Tour*, New York, NY: Oxford University Press.

Monge, Peter R., and Noshir Contractor (1993) *Theories of Communication Network*, New York, NY: Oxford University Press.

Sawyer, Keith, R. (2005) *Social Emergence, Societies as Complex Systems*, New York, NY: Cambridge University Press.

Sawyer, R. Keith (2008) *Group Creativity: Music, Theater, Collaboration*, Mahwah, NJ: Lawrence Erlbaum Associates.

Schroeder, Manfred (2009) *Fractals, Chaos, Power Law, Minutes from an Infinite Paradise*, Mineola, NY: Dover Publications, Inc.

Schwartz, Glen M., and J. Nichols, eds. *After Collapse: The Regeneration of Complex Societies*, Tucson, AZ: The University of Arizona Press.

Skyrms, Brian (2014) *Social Dynamics*, New York, NY: Oxford University Press.

Smith, Adam (2013) *The Wealth of Nations*, Scotts Valley, CA: CreateSpace Independent Publishing Platform.

Theise, Neil (2023) *Notes on Complexity: A Scientific Theory of Connection, Consciousness, and Being*, New York, NY: Spiegel and Grau.

Waldrop, M. Mitchell (1993) *Complexity: The Emerging Science at the Edge Order and Chaos*, New York, NY: Open Road.

West, Geoffrey (2018) *Scale: The Universal Laws of Life, Growth, and Death in Organisms, Cities, and Companies*, New York, NY: Penguin.

INDEX